Lecture Notes in Mathematics

2175

Editors-in-Chief:
J.-M. Morel, Cachan
B. Teissier, Paris

Advisory Board:
Camillo De Lellis, Zurich
Mario di Bernardo, Bristol
Michel Brion, Grenoble
Alessio Figalli, Zurich
Davar Khoshnevisan, Salt Lake City
Ioannis Kontoyiannis, Athens
Gabor Lugosi, Barcelona
Mark Podolskij, Aarhus
Sylvia Serfaty, New York
Anna Wienhard, Heidelberg

More information about this series at http://www.springer.com/series/304

Saint-Flour Probability Summer School

The Saint-Flour volumes are reflections of the courses given at the Saint-Flour Probability Summer School. Founded in 1971, this school is organised every year by the Laboratoire de Mathématiques (CNRS and Université Blaise Pascal, Clermont-Ferrand, France). It is intended for PhD students, teachers and researchers who are interested in probability theory, statistics, and in their applications.

The duration of each school is 13 days (it was 17 days up to 2005), and up to 70 participants can attend it. The aim is to provide, in three high-level courses, a comprehensive study of some fields in probability theory or Statistics. The lecturers are chosen by an international scientific board. The participants themselves also have the opportunity to give short lectures about their research work.

Participants are lodged and work in the same building, a former seminary built in the 18th century in the city of Saint-Flour, at an altitude of 900 m. The pleasant surroundings facilitate scientific discussion and exchange.

The Saint-Flour Probability Summer School is supported by:
- Université Blaise Pascal
- Centre National de la Recherche Scientifique (C.N.R.S.)
- Ministère délégué à l'Enseignement supérieur et à la Recherche

For more information, see

http://recherche.math.univ-bpclermont.fr/stflour/stflour-en.php

Christophe Bahadoran
bahadora@math.univ-bpclermont.fr

Arnaud Guillin
Arnaud.Guillin@math.univ-bpclermont.fr

Laurent Serlet
Laurent.Serlet@math.univ-bpclermont.fr

Université Blaise Pascal – Aubière cedex, France

Francis Comets

Directed Polymers in Random Environments

École d'Été de Probabilités de Saint-Flour
XLVI – 2016

 Springer

Francis Comets
Mathematics, case 7012
Université Paris Diderot - Paris 7
Paris, France

ISSN 0075-8434   ISSN 1617-9692   (electronic)
Lecture Notes in Mathematics
ISBN 978-3-319-50486-5   ISBN 978-3-319-50487-2   (eBook)
DOI 10.1007/978-3-319-50487-2

Library of Congress Control Number: 2017931071

Mathematics Subject Classification (2010): Primary: 60K37; secondary: 82B44, 60J10, 60H05, 60F10

Printed on acid-free paper

This Springer imprint is published by Springer Nature
The registered company is Springer International Publishing AG
The registered company address is: Gewerbestrasse 11, 6330 Cham, Switzerland

L'ordre est le plaisir de la raison,
mais le désordre
est le délice de l'imagination.

Paul Claudel,
Introduction of "Le soulier de satin" (1924)

Preface

This monograph contains the notes of lectures I gave in Saint Flour Probability Summer School in July 2016. The two other courses were given by Paul Bourgade and by Scott Sheffield and Jason Miller. The material was grown in several courses over the last 15 years in the Universities of Paris Diderot and Campinas and also at CIRM in Marseille, Max Planck Institute in Leipzig, and at the ALEA Conference.

The model of directed polymers in random environment is a simplified model for stretched elastic chains which are pinned by random impurities. The main question is:

> What does a random walk path look like if rewards and penalties are randomly distributed in the space?

In large generality, it experiences a phase transition from diffusive behavior to localized behavior. In this monograph, we place a particular emphasis on the localization phenomenon. The model can be mapped, or it relates, to many other ones, including interacting particle systems, percolation, queueing systems, randomly growing surfaces, and biological population dynamics. Even though it attracts a huge research activity in stochastic processes and in statistical physics, it still keeps many secrets, especially in space dimension two and larger. It has non-Gaussian scaling limits, and it belongs to the so-called KPZ universality class when the space dimension is one. We adopt a statistical mechanics (Gibbsian) approach, using general and powerful tools from probability theory, to successfully tackle the general framework. To obtain precise asymptotics, we also study a particular model (the so-called log-gamma polymer) which is exactly solvable.

We consider the simplest discrete setup all through book. The only exception is the KPZ equation—a stochastic partial differential equation—for the decisive reason that it is the scaling limit of the discrete model in one dimension with vanishing interaction, and therefore, it finds a natural place. I have chosen not to include material covering other continuous or semi-discrete versions of the model; this is the reason why I do not present all available techniques here. I hope that my choice to stick to the simplest setup makes the purpose crystal clear and the arguments transparent.

Giving the state of art from different perspectives, the monograph is devised for researchers interested in polymer models. Written in the format of a first course on the subject, it is also accessible to master's and Ph.D. students in probability or statistical physics. In this perspective, I have deliberately preferred simple proofs based on first principles and general methods rather than optimal results requiring technical and specific proofs. The mathematician reader will understand how to extend the proof to a wider framework and to get optimal constant. Researchers being nowadays incented for efficiency reasons to be more and more specialized in narrow subjects, I hereby advocate for wide view on the field, fundamental objects, and robust techniques. The whole text can be read linearly, and technicalities appear in a progressive manner.

I was a junior mathematician in the mid-1980s when Erwin Bolthausen visited Orsay University, just after he brought polymer models from statistical physics into the realm of probability using martingale techniques in his paper [46]. He was sharing his enthusiasm, and indeed the challenge looked great. Since the beginning of our century, the model has attracted many different methods from various fields of mathematics including orthogonal polynomials, random matrices, semimartingales, stochastic PDEs, integrable systems, tropical combinatorics, and representation theory. It is a challenging experience for me to write these notes, which cover only a part of the picture. I first came to Saint Flour Probability Summer School in 1981 as a student and, also many times since that time, found the format suitable for learning new topics in the monastic atmosphere of the Grand Séminaire. I warmly thank Christophe Bahadoran, Arnaud Guillin, and Laurent Serlet for organizing the school in a friendly and enjoyable ambiance, pursuing a long-standing, notorious tradition of training in probability, and giving me a definitive reason to write these notes.

I am first of all indebted in exchanges on this particular subject with wonderful collaborators over the years: Mike Cranston, Ryoki Fukushima, Quansheng Liu, Shuta Nakajima, Vu-Lan Nguyen, Sergei Popov, Jeremy Quastel, Alejandro Ramírez, Tokuzo Shiga, Marina Vachkovskaia, Vincent Vargas, and Nobuo Yoshida. The works of many colleagues have had a profound impact on my research and view of the subject, in particular Anton Bovier, Ivan Corwin, Bernard Derrida, Frank den Hollander, Pierre Le Doussal, Kostya Khanin, Firas Rassoul-Agha, Timo Seppäläinen, and Herbert Spohn. I acknowledge KITP for its hospitality during the program on KPZ integrability and universality, which have been a great update for me to write Chap. 7. I owe special thanks to Quentin Berger, Chris Janjigian, Oren Louidor, Gregorio Moreno, Makoto Nakashima, and Nikos Zygouras for correcting parts of the manuscript and suggesting improvements. The notes have benefited from further exchanges with Ronfeng Sun and Attila Yilmaz. Questions and comments by the participants, including Reda Chhaibi, Aser Cortines, Clément Cosco, and Giambattista Giacomin, helped me to improve the matter. I apologize for the (numerous) remaining errors; they are mine!

Paris, France   Francis Comets
October 2016

Contents

List of Symbols

This list indicates the page number where the main symbols are defined:

Greek symbols

Roman symbols

Calligraphic

Other

Chapter 1
Introduction

1.1 Polymer Models

The model we consider all through these notes is easy to define as a random walk in a random potential.

1.1.1 Random Walk in Random Environment: A Model for Directed Polymers

First fix a few notations.

- *The random walk:* $(S = \{S_n\}_{n\geq 0}, P_x)$ is a simple random walk on the d-dimensional integer lattice $\mathbb{Z}^d$ starting from $x \in \mathbb{Z}^d$. Precisely, the random sequence S is defined on the probability space $\Omega_{\text{traj}} = (\mathbb{Z}^d)^{\mathbb{N}}$ with the cylindric σ-field $\mathscr{F}$ and a probability measure P_x such that, under P_x, the jumps $S_1 - S_0, \ldots, S_n - S_{n-1}$ are independent with

$$P_x\{S_0 = x\} = 1, \quad P_x\{S_n - S_{n-1} = \pm e_j\} = (2d)^{-1}, \quad j = 1, 2, \ldots, d,$$

where $e_j = (\delta_{kj})_{k=1}^d$ is the jth vector of the canonical basis of $\mathbb{Z}^d$. In the sequel, $P_x[X]$ denotes the P_x-expectation of a r.v. (random variable) X, and P_0 will be simply written by P.
- *The random environment:* $\omega = \{\omega(n, x) : n \in \mathbb{N}, x \in \mathbb{Z}^d\}$ is a sequence of r.v.'s which are real valued, non-constant, and i.i.d. (independent identically distributed) r.v.'s defined on a probability space $(\Omega, \mathscr{G}, \mathbb{P})$ such that

$$\mathbb{P}[\exp(\beta\omega(n, x))] < \infty \quad \text{for all } \beta \in \mathbb{R}. \tag{1.1}$$

© Springer International Publishing AG 2017
F. Comets, *Directed Polymers in Random Environments*, Lecture Notes
in Mathematics 2175, DOI 10.1007/978-3-319-50487-2_1

Here, and in the sequel, $\mathbb{P}[Y]$ the $\mathbb{P}$-expectation of a random variable Y defined on $(\Omega, \mathscr{A}, \mathbb{P})$ and $\mathbb{P}[Y; A]$ the $\mathbb{P}$-expectation of Y on the event $A \in \mathscr{A}$. We will take $\Omega = \mathbb{R}^{\mathbb{N}^* \times \mathbb{Z}^d}$ the canonical space for definiteness.

From these two basic ingredients we define the object we consider in the notes.

- *The polymer measure:* For any $n > 0$, define the probability measure $P_n^{\beta,\omega}$ on the path space $(\Omega_{\text{traj}}, \mathscr{F})$ by

$$P_n^{\beta,\omega}(d\mathbf{x}) = \frac{1}{Z_n(\omega, \beta)} \exp\{\beta H_n(\mathbf{x})\}\, P(d\mathbf{x}), \tag{1.2}$$

where $\beta > 0$ is a parameter (the inverse temperature), where

$$H_n(\mathbf{x}) = H_n^{\omega}(\mathbf{x}) = \sum_{1 \leq j \leq n} \omega(j, x_j) \tag{1.3}$$

is the energy of the path $\mathbf{x}$ in environment ω (Hamiltonian potential) and

$$Z_n = Z_n(\omega, \beta) = P\left[\exp\left(\beta \sum_{1 \leq j \leq n} \omega(j, S_j)\right)\right] \tag{1.4}$$

is the normalizing constant to make $P_n^{\beta,\omega}$ a probability measure. From its definition (1.2), $P_n^{\beta,\omega}$ is the Gibbs measure with Boltzmann weight $\exp\{\beta H_n\}$, and Z_n is the so-called partition function. Of course, in the present context, the above expectation is simply a finite sum,

$$Z_n(\omega, \beta) = \sum_{\mathbf{x}} (2d)^{-n} \exp(\beta H_n(\mathbf{x}))$$

where $\mathbf{x}$ ranges over the $(2d)^n$ possible paths of length n for the simple random walk.

The polymer measure $P_n^{\beta,\omega}$ can be thought of as a Gibbs measure on the path space $(\Omega, \mathscr{F})$ with the Hamiltonian H_n. We stress that the random environment ω is contained in both $Z_n(\omega, \beta)$ and $P_n^{\beta,\omega}$ without being integrated out, so that they are r.v.'s on the probability space $(\Omega, \mathscr{G}, \mathbb{P})$. The polymer is attracted to sites where the random environment is positive, and repelled by sites where the environment is negative.

1.1.2 Modelization: Polymer in an Emulsion with Repulsive Impurities

We start with an informal description of a specific example. Consider a hydrophilic polymer chain (i.e., a long chain of monomers) wafting in water. Due to the thermal fluctuation, the shape of the polymer should be understood as a random object. We now suppose that the water contains randomly placed hydrophobic molecules as impurities, which repel the hydrophilic monomers which the polymer consists of. The question we address here is:

How does the impurities affect the global shape of the polymer chain?

We try to answer this question in a mathematical framework. However, as is everywhere else in mathematical physics, it is very difficult to do so without compromising with a rather simplified picture of the initial problem. Here, our simplification goes as follows. We first suppress entanglement, self-intersections and U-turns of the polymer; we shall represent the polymer chain as a graph $\{(j, x_j)\}_{j=1}^n$ in $\mathbb{N} \times \mathbb{Z}^d$, so that the polymer is supposed to live in $(1 + d)$-dimensional discrete lattice and to stretch in the direction of the first coordinate. Such a model is called *directed*. Each point $(j, x_j) \in \mathbb{N} \times \mathbb{Z}^d$ on the graph stands for the position of jth monomer in this picture. Secondly, we assume that, the transversal motion $\mathbf{x} = \{x_j\}_{j=1}^n$ performs a simple random walk in $\mathbb{Z}^d$, if the impurities are absent. This accounts for consecutive monomers in the chain taking all possible configurations at a fixed distance one from another. We then define the energy of the path $\{(j, x_j)\}_{j=1}^n$ by the formula (1.3) at some (inverse) temperature $\beta > 0$, where $\{\omega(n, x) : n \geq 1, \ x \in \mathbb{Z}^d\}$ is a field of real i.i.d. random variables, with $\omega(n, x)$ describing the presence (or strength) of an impurity at site (n, x). The case of an emulsion announced in the title is well captured by an i.i.d. model. Every time the path steps on an impurity it gets a penalty, and every step on a water molecule brings a reward. The typical shape $\{(j, x_j)\}_{j=1}^n$ of the polymer is then given by the one that maximizes the energy (1.3). Let us suppose for example that $\omega(n, x)$ takes two different values $+1$ ("presence of a water molecule at (n, x)") and -1 ("presence of the hydrophobic impurity at (n, x)"). Then, the energy of the polymer is decreased by β each time a monomer is in contact with the impurity ($\omega(j, x_j) = -1$). Therefore, the typical shape of the polymer for each given configuration of $\{\omega(j, x)\}$ is given by the one which tries to avoid the impurities as much as possible.

What Are the Main Questions? We want to study large systems, that is, asymptotics when the length of the polymer chain goes to infinity, for typical realizations of ω.

Mechanically, the elasticity of the polymer competes with the energy of very favorable, but distant, pinning sites that would require a costly distortion of the polymer. From this competition we can guess two possible scenarios:

1. If the space dimension is large and the temperature is high, the impurities should not affect the global shape of the polymer. This is because the polymer has much space to avoid the impurities by the first assumption and the impurities are weak enough by the second one. It has no interest in travelling far, it only wants to pick a specific proportion (a little bigger than the density) of rewards, using a local strategy.
2. If the dimension is small or the environment is strong, then the polymer will not be able any more to avoid the impurities if it stays at reasonable distances. The path will have an advantage in travelling far in order to find more favourable areas. These atypical areas in the medium correspond to a higher density of rewards, and a precise geometry making them feasible for the walk. Then, the global shape of the polymer path changes in a drastic manner.

The informal description given above has been put into a mathematical framework by the formalism of a Gibbs measure, in the framework of statistical mechanics. For a more substantial overview of polymer modelization, see the first chapter of den Hollander [92].

1.1.3 Polymers Viewed as Percolation at a Positive Temperature

For a fixed n, we can take the zero-temperature limits: as $\beta \to \infty$,

$$\begin{cases} \beta^{-1} \ln Z_n(\omega, \beta) & \longrightarrow & \max_{\mathbf{x}} H_n(\mathbf{x}), \\ P_n^{\beta, \omega}(\cdot) & \text{concentrates on } \arg\max_{\mathbf{x}} H_n(\mathbf{x}). \end{cases} \tag{1.5}$$

The right-hand sides of (1.5) are respectively the passage time and the set of geodesics of a last passage percolation problem. Such quantities have been extensively studied [102, 153, 186], and the last section of Auffinger et al. [20] for a recent account. Percolation here is oriented, site percolation. Thus, the polymer measure is an interpolation between simple random walk (corresponding to $\beta = 0$) and geodesics in last passage percolation (limit as $\beta \to \infty$). In particular it can be viewed as a "fuzzy" geodesics, and it allows to study percolation by decreasing continuously the temperature, performing a perturbative approach starting from the (well-known) random walk model. Having the parameter β at hand will reveal most useful.

1.1.4 Exponents and Localization

To quantify how the fluctuations are influenced by the disorder, two **characteristic exponents**, $\chi^{\|} \in [0, 1/2]$ and $\chi^{\perp} \in [1/2, 1]$, are introduced: loosely speaking, they are defined by

$$P_n^{\beta,\omega}\left[|S_n|^2\right] \approx n^{2\chi^{\perp}}, \quad \mathrm{Var}\left(\ln Z_n(\omega, \beta)\right) \approx n^{2\chi^{\|}}. \tag{1.6}$$

The volume exponent $\chi^{\|}$ characterizes the fluctuations of thermodynamic quantities, though the wandering exponent (or roughness exponent) characterizes the transverse fluctuations of the path. It is expected, and partially confirmed (see [16] adapting ideas for last passage percolation summarized in [63]) that the exponents are related via

$$\chi^{\|} = 2\chi^{\perp} - 1. \tag{1.7}$$

The scenario (1) in Sect. 1.1.2 above corresponds to diffusive behavior $\chi^{\perp} = 1/2$, though case (2) has $\chi^{\perp} > 1/2$ (superdiffusivity). In scenario (2) these exponents are expected to be universal, i.e., not depending on the details of the model but only on the dimension, with $\chi^{\|} = 1/3$, $\chi^{\perp} = 2/3$ for $d = 1$.

Another feature of scenario (1) is **localization**: the polymer has a large probability to lie in a favourable regions with high density of rewards and feasible for the path: we call them corridors, and this phenomenon is called path localization. It will be studied in Chap. 5 in a general framework and Chap. 7 on a specific integrable model, where explicit computations can be performed.

The interest of mathematicians for these characteristic exponents, their relation and some inequalities depending on the space dimension, started two decades ago for last passage percolation in the discrete setting [164, 187, 194] and in the continuous setting [240, 241]; it remains very strong with novel contributions [17, 63]. For polymers, see also [35, 59, 73, 171, 193] for various estimates in the general setup. But except for exactly solvable models in $d = 1$, they remain most mysterious—particularly for $d \geq 2$—in spite of many efforts.

1.2 Particle in a Random Potential

We relate the Gibbs measure to branching random walks in an inhomogeneous medium.

1.2.1 Particle in Deadly Obstacles

In order to give a simple intuition of what is a Gibbs measure, we first discuss here our model in the case when the environment variables are uniformly bounded from above. For simplicity, and without loss of generality,[1] we assume in this section that

$$\omega(t, x) \leq 0 \qquad \mathbb{P} - \text{a.s.} \tag{1.8}$$

In this case, we can interpret the model in terms of a walker moving among deadly obstacles. The space location x hosts a deadly obstacle which strength at time t is $(-\beta\omega(t, x))$. We start with an informal description. A particle starts at the origin at time 0, it jumps at integer times and immediately after, it dies or survives. It moves according to a simple random walk when alive, but it has a probability $\exp \beta\omega(t, x) \in (0, 1]$ to survive the obstacle at time t if it is still alive at time $t-1$ and has jumped at location x at time t, and complementary probability $1 - \exp \beta\omega(t, x)$ to be killed by the obstacle. The event of dying or surviving at time t are independent of the past, conditionally on being alive at time $t - 1$ and on the step at time t.

We now give a precise construction of this process. Let $P, \mathbb{P}$ as before and let Unif be the uniform law on $[0, 1]$. We fix a realization $(\omega(t, x); (t, x) \in \mathbb{N} \times \mathbb{Z}^d)$ from $\mathbb{P}$. On the product space $\Omega_{\text{traj}} \times [0, 1]$, we define the product measure $Pr = P \otimes \text{Unif}$, and denote by $U : (\mathbf{x}, u) \mapsto u$ the second coordinate mapping. Let

$$\tau = \inf \left\{ n \geq 1 : U > \exp\{\beta H_n^\omega(S)\} \right\}$$

be the lifetime of the path S. We also add to $\mathbb{Z}^d$ a special state called cemetery point and denoted by $\dagger$. Then, the process defined by

$$X_n = \begin{cases} S_n & \text{if } n < \tau, \\ \dagger & \text{if } n \geq \tau, \end{cases}$$

corresponds to our informal description, and is intimately related to the polymer path (Fig. 1.1).

Proposition 1.1 *The sequence X is a (time-inhomogeneous) Markov chain on $\mathbb{Z}^d \cup \{\dagger\}$. It has $\dagger$ as absorbing state, and its one-step transition probability from $x \in \mathbb{Z}^d$ is given by*

$$Pr(X_{n+1} = y | X_n = x) = (2d)^{-1} \exp\{\beta\omega(n + 1, y)\} \qquad \text{if } y \in \mathbb{Z}^d,$$

---

[1] A constant shift for $\omega(t, x)$ has a trivial effect on the objects we consider.

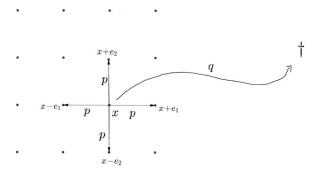

Fig. 1.1 A particle in a deadly potential: The state space is $\mathbb{Z}^d \cup \{\dagger\}$, with $d = 2$ and $\dagger$ is a cemetery (absorbing) point. Conditionally on being still alive at time t and at location x, it dies out there with probability $q = 1 - \exp \beta \omega(t, x)$ (recall that $\omega \leq 0$), or survives one more time unit at least and moves to a neighboring site with probability $p = (2d)^{-1}(1 - q)$. This description corresponds to a slightly different potential $\tilde{H}_n(S) = \sum_{i=0}^{n-1} \omega(i, S_i)$

and

$$Pr(X_{n+1} = \dagger | X_n = x) = (2d)^{-1} \sum_{z \in \mathbb{Z}^d, |z-x|_1 = 1} \left(1 - \exp\{\beta\omega(n+1, z)\}\right).$$

Finally, the partition function of the polymer model is equal to the survival probability,

$$Pr(\tau > n) = Z_n(\omega, \beta),$$

and the polymer measure with time horizon n is equal to the law of the path of the particle killed in the random potential conditioned to be alive at time n,

$$P_n^{\beta, \omega}(\mathbf{x}) = Pr(X_{[0,n]} = \mathbf{x} | \tau > n) . \tag{1.9}$$

Proof First, we note that, for all n and all path $\mathbf{x}$ of length n,

$$Pr(\tau > n | S_{[0,n]} = \mathbf{x}) = \text{Unif}(U \leq \exp\{\beta H_n(\mathbf{x})\})$$
$$= \exp\{\beta H_n(\mathbf{x})\},$$

and that

$$Pr(\tau > n, S_{[0,n]} = \mathbf{x}) = (2d)^{-n} \exp\{\beta H_n(\mathbf{x})\}, \tag{1.10}$$

and also that

$$Pr(\tau > n) = Pr\big[Pr(\tau > n | S_{[0,n]})\big]$$
$$= Z_n(\omega, \beta).$$

Therefore,

$$Pr(X_{[0,n]} = \mathbf{x} | \tau > n) = Pr(S_{[0,n]} = \mathbf{x} | \tau > n)$$
$$= \frac{Pr(\tau > n | S_{[0,n]} = \mathbf{x}) Pr(S_{[0,n]} = \mathbf{x})}{Pr(\tau > n)}$$
$$= P_n^{\beta,\omega}(\mathbf{x}),$$

which is (1.9). We finally check that the dynamics of the random sequence $X = (X_n, n \geq 0)$ corresponds to the description at the beginning of the section. First, X is a Markov chain on $\mathbb{Z}^d \cup \{\dagger\}$, with $\dagger$ as absorbing state. Moreover, for all nearest-neighbor path ω in $\mathbb{Z}^d$ with $\mathbf{x}_n = x, \mathbf{x}_{n+1} = y$,

$$Pr\left(\tau > n + 1, S_{[0,n+1]} = \mathbf{x}_{[0,n+1]} \mid \tau > n, S_{[0,n]} = \mathbf{x}_{[0,n]}\right)$$
$$= \frac{Pr(\tau > n + 1, S_{[0,n+1]} = \mathbf{x}_{[0,n+1]})}{Pr(\tau > n, S_{[0,n]} = \mathbf{x}_{[0,n]})}$$
$$\overset{(1.10)}{=} (2d)^{-1} \exp\{\beta(H_{n+1}(\mathbf{x}) - H_n(\mathbf{x}))\}$$
$$= (2d)^{-1} \exp\{\beta\omega(n + 1, y)\}.$$

This implies the first two equalities of the proposition. □

1.2.2 Branching Random Walk Picture

We discuss now the case when the environment takes only nonnegative values. We can interpret the model in terms of a branching population (without deaths) moving in a random potential: The polymer measure with time horizon n is equal to the law of the ancestral path of an individual selected uniformly at random in the full population at time t. Moreover, the partition function $Z_{n,\beta}^{\omega}$ of the polymer model is equal to the expected population size at time n in a fixed environment ω.

The case of general environment is a simple combination of the two above: one can define a branching random walk in random environment such that, conditionally on survival at time n, the law of the ancestral path of an individual selected uniformly at random in the full population at that time is the polymer measure [76].

All these representations are in the spirit of Feynman-Kac formula.

1.3 History, Experiments and Related Models

The polymer model was originally introduced in physics literature [140] to mimic the phase boundary of Ising model subject to random impurities. It became a popular model for kinetic roughening of growing surfaces including front propagation, bacterial colonies, vortex lines in supraconductors, ... [156]. Also, it has inspired a number of methods for its analysis (replica trick, renormalization group theory, Bethe ansatz, scaling theory). The model soon reached the mathematics community [46, 141], where it was explicitly formulated as the polymer measure (1.2).

Let us mention some physical realizations and real experiments fitting the model.

1. Torn paper sheets [152]. A rectangular sheet of paper is tightened by a machine. The strain is applied on two opposite sides, and it is slowly increased. A small notch is made on a third side to initiate the tear. The fracture is governed by the random geometry of the fiber network. The fracture line is highly correlated to the weakest bonds in the sheet, as can be checked by microdensitometry. Directed polymers with $d = 1$ and its zero temperature counterpart, oriented first passage percolation, are natural model for the fracture line.
2. Nematic crystal: Takeuchi and Sano [224] realized an real experiment of a growing domain of topological-defect turbulence in electrically driven liquid crystal. This gives an experimental evidence for KPZ scaling and Tracy-Widom fluctuations. Two videos of a flat or circular interface are available, see
   https://www.youtube.com/watch?v=tcdufma7UJA
   https://www.youtube.com/watch?v=useN8ZdRvb8.

The polymer model has a continuous limit expressed by a stochastic partial differential equation, the so-called Kardar-Parisi-Zhang equation (8.1), which is the generic model for a variety of random growth phenomena and covers the statistics of roughness. The corresponding universality class consists in models which large-scale behavior share the same statistics. In spite of being 30 years old, KPZ equation remains of a tremendous interest in physics, as shown in recent accounts [132] and [229].

As mentioned above, corresponding to $\beta = \infty$, the ground states—i.e. the paths $\mathbf{x}$ maximizing H_n—are the geodesics of last passage percolation. The relations of polymers to last passage percolation, extends to a number of related models: tandem queueing systems, totally asymmetric exclusion process, edge of the spectrum of random matrices, ...

The parabolic Anderson model is a time-continuous version of our discrete model, it is a popular example for intermittency phenomenon [61]. It has been studied in many facets: fine asymptotics for vanishing diffusivity [62, 87], large deviations showing the asymmetry [88], using Gaussian tools [53]. A complete review is [115] in the time-independent case. An intermediate case is the so-called random walk in a dynamic random environment [106].

To complete the section, we indicate a few models which are different but closely related:

1. Polymer pinning: in which localization only occurs at a known, specific location if it does occur. Main example: defect lines in the plane where rewards are available (and no rewards elsewhere). Rewards may be deterministic or random. See [92, 119, 120] for surveys of a huge literature.

2. Brownian polymers in Poissonian medium: Here, the polymer is given by a Brownian motion, whereas the environment is given by a space-time Poisson point process which is enlarged in the transverse directions by a fixed radius. All our forthcoming results have a counterpart here [73, 74, 234]. The model considered here enables to use stochastic calculus, with respect to both Brownian motion and Poisson process, leading to handy formulas for fluctuations analysis and qualitative properties of the phase diagram. The relation to some formulation of the Kardar-Parisi-Zhang equation is transparent. Girsanov formula and the invariance of the Poisson process yields an explicit rate function for large deviation of the path [74] and integration by parts yield complete localization [77].

3. In a similar spirit, it is natural to model the path itself by an independent increment process. Lévy polymers in random environment are considered in [68, 173]. Due to long jumps, the critical dimension can be different from that of diffusive models.

4. Random walks in a *time-independent* potential: spectral methods [222, 223] are powerful in this case. The connection with our setup is when considering crossings of such walks when the path is stretched to a ballistic behaviour (bridge) [142, 222]. Backtracks of the path are penalized, and the model we consider in this book becomes a reasonable approximation. In fact it corresponds to forbid backtracks and even more, to have a systematic move at each step in a fixed direction.

5. The model here fits the paradigm of disordered systems and random medium: local inhomogeneities with large scale regularity [52] are nicely approximated by a random medium.

6. As mentioned in Sect. 1.2.2 there is a strong connection with branching random walks in random environment. For a better focus we will not develop these models. We simply mention here that, analogous to the results we discuss here, diffusive regime [136, 213, 243] and localized behavior [138, 214] have been studied in details.

7. Linear stochastic evolution is a class of stochastic growth models on the integer lattice $\mathbb{Z}^d$, including a time discretization of the "linear systems" discussed in [165]. Examples of models in this class are the number of open paths in oriented percolation and the binary contact path process. The counterparts of the topics we study here can be found in [112, 183, 244].

8. The effect of the correlated noise has been studied under various aspects: space-correlation in the discrete case [158] and in the continuous case [36, 149], time-correlation in the continuous case [236].

9. Localization covers the absence of diffusion in random medium: inhomogeneities can pin random paths and cancel their natural diffusive behavior. Anderson's original 1958 paper addressed the question of spin diffusion on the lattice and electrical conduction in a semiconductor impurity band. The phenomenon has received a huge interest to understand its mechanism as well as for its applications in science and material engineering. We refer to the reviews [160, 230] for a physics account, and [139] for a mathematical introduction to localization in Anderson model.

1.4 Highlights

Accessible and building on general methods, the monograph is designed to be read linearly. The main topics are organized as described below.

In Chap. 2 we take the view point of statistical mechanics for disordered systems. We introduce the free energy, prove self-averaging, determine the qualitative shape the phase diagram and find a thermodynamic phase transition on the basis of a comparison with an annealed model. As stated in Theorem 2.4, the critical temperature β_c is determined by the onset of strict inequality between the quenched and annealed rate of growth of the partition function,

$$\beta > \beta_c \iff \lim_n n^{-1} \ln Z_n(\omega, \beta) < \lim_n n^{-1} \ln \mathbb{P}Z_n(\omega, \beta), \qquad (1.11)$$

$$0 \le \beta \le \beta_c \iff \lim_n n^{-1} \ln Z_n(\omega, \beta) = \lim_n n^{-1} \ln \mathbb{P}Z_n(\omega, \beta). \qquad (1.12)$$

In Chap. 3 we take a more probabilistic approach, showing that martingale theory yields direct information on the rescaled partition function. For instance $W_n = Z_n(\omega, \beta)/\mathbb{P}[Z_n(\omega, \beta)]$ is a positive, mean-one martingale. The well-known "second moment method" reveals efficient in a large regime, where analytic methods can be used. Computing integer moments, we will check that the martingale W_n is bounded in L^2 when d is large and β is small, implying that the phase transition occurs in a non-trivial manner in dimension 3 and above (cf. Theorem 3.3). Sharp results can be obtained in this region, such as the existence of a finite random limit for $P_n^{\beta,\omega}[|S_n|^2] - n$ or $P_n^{\beta,\omega}[H_n] - n\lambda'(\beta)$.

In Chap. 4 we introduce the polymer models on trees—or, equivalently, the branching random walk in random environment, which can be seen as a mean-field approximation. This model is solvable, and we will discuss similarities and differences with the polymer on the lattice.

In Chap. 5 we use semi-martingale techniques for a fine analysis of the weak and the strong disorder regimes, relating these regimes to the fine behavior of the random polymer. Standard tools from stochastic processes, as quadratic variations, yield quantitative information on the overlap between replicas, i.e., independent copies of the polymer path sharing the same environment. This is the key to understanding that the thermodynamic phase transition coincide with the localization transition. We will obtain in Theorem 5.4 that $\beta > \beta_c$ is equivalent to the positivity of the Cesaro limit of the probability of the favorite end-point of the polymer,

$$J_n = \max\{P_n^{\beta,\omega}(S_n = x); x \in \mathbb{Z}^d\}.$$

Thus, the critical temperature is the onset of localization of the end-point of the polymer.

In Chap. 6 we study the localized phase. It happens at all temperature in dimension 1 and 2. We describe the corridors where the polymer concentrates. The last section will deal with the simpler situation where the medium has heavy tails. Then, a coarse graining is performed to analyze localization.

In Chap. 7 we study in details the log-gamma polymer recently introduced by Seppäläinen, which is integrable in one dimension. Explicit computations are possible, and we will review some of the precise results from an abundant literature, including characteristic exponents and fluctuation statistics. Slightly modifying the model on the boundaries we will have the benefit of stationarity, and we will completely determine the limiting distribution of J_n, cf. Corollary 7.3. Also, in this model with boundaries, we will obtain tightness of the polymer around the favorite endpoint, and the large scale fluctuation of the favorite endpoint. The mechanism for localizing the polymer is by confinement in the deepest valley a random potential— here, a random walk. We believe that this mechanism, which appears in Sinai's localization for the random walk in random environment in the one-dimensional recurrent case [215], is an important ingredient of polymer localization in general.

Chapter 8 will be dedicated to the KPZ equation in dimension $d = 1$. We will discuss what is the meaning of the equation, express its solution as a series, define the continuum random polymer that it generates, and study its relation to intermediate disorder for polymers. We will show convergence of the rescaled, discrete polymer in dimension $d = 1$ and with a suitably decaying inverse temperature $\beta = \beta_n$, to the continuum random polymer.

In Chap. 9 we will discuss the variational formulas giving the value of free energy. They come in two forms: optimizing over cocycles as in stochastic homogenization, or optimizing over probability measures as in usual Gibbs variational formulas, giving a balance between energy and entropy.

Finally, an appendix collects a few general tools—mainly from stochastic processes and statistical mechanics—which are used all through the monograph.

Chapter 2
Thermodynamics and Phase Transition

An important quantity for this model is the logarithmic moment generating function λ of $\omega(n, x)$,

$$\lambda(\beta) = \ln \mathbb{P}[\exp(\beta\omega(n, x))], \qquad (2.1)$$

which is finite for all real β by assumption (1.1). For example, $\lambda(\beta) = \ln(pe^{-\beta} + (1-p)e^{\beta})$ for the Bernoulli environment and $\lambda(\beta) = \frac{1}{2}\beta^2$ for the Gaussian environment. All through these notes, we will restrict to

$$\boxed{\beta \geq 0}.$$

Doing this we do not lose generality, since the opposite case being obtained by considering environment $-\omega$.

2.1 Preliminaries

On the space Ω of environments, define for $i \geq 1, x \in \mathbb{Z}^d$, the **shift operator** $\theta_{i,x} : \Omega = \mathbb{R}^{\mathbb{N}^* \times \mathbb{Z}^d} \to \Omega$ given by $\omega \mapsto \theta_{i,x}\omega$,

$$(\theta_{i,x}\omega)(t, y) = \omega(i + t, x + y) . \qquad (2.2)$$

Thus, $\theta_{i,x}\omega$ is simply the field of environment variables which is seen from the "point" (i, x). Its law is the same as the law of ω itself, i.e., the product law $\mathbb{P}$.

© Springer International Publishing AG 2017
F. Comets, *Directed Polymers in Random Environments*, Lecture Notes
in Mathematics 2175, DOI 10.1007/978-3-319-50487-2_2

2.1.1 Markov Property and the Partition Function

For $n, m \geq 1$, $x \in \mathbb{Z}^d$, the random variable

$$Z_m \circ \theta_{n,x}(\omega) = Z_m(\theta_{n,x}\omega; \beta) = P_x\left[\exp\left(\sum_{1 \leq t \leq m} \beta\omega(t+n, S_t)\right)\right],$$

is the partition function of the polymer of length m starting at x at time n. Since ω and its shift $\theta_{n,x}\omega$ have the same law, $Z_m \circ \theta_{n,x}$ has the same law as Z_m. By the Markov property of the random walk, we can also write it in the form of a conditional expectation given $\mathscr{F}_n = \sigma\{S_t, t \leq n\}$,

$$Z_m \circ \theta_{n,x}(\omega) = P\left[e^{\beta\{H_{n+m}(S)-H_n(S)\}}|\mathscr{F}_n\right] \quad \text{on the event } \{S_n = x\}.$$

For $n, m \geq 1$, we can express the partition function of the polymer of length $n + m$ by conditioning:

$$\begin{aligned}
Z_{n+m} &= P\left[e^{\beta H_{n+m}(S)}\right] \\
&= P\left[e^{\beta H_n(S)}P\left[e^{\beta\{H_{n+m}(S)-H_n(S)\}}|\mathscr{F}_n\right]\right] \\
&= P\left[e^{\beta H_n(S)} \times Z_m \circ \theta_{n,S_n}\right]
\end{aligned}$$

using the previous observation. This important identity will be referred to as the Markov property. It can be reformulated as

$$Z_{n+m} = Z_n \times P_n^{\beta,\omega}\left[Z_m \circ \theta_{n,S_n}\right]. \tag{2.3}$$

2.1.2 The Polymer Measure as a Markov Chain

In this section we discuss some basic properties of the polymer measure $P_n^{\beta,\omega}$.

Let us fix the environment ω. Then, under the polymer measure $P_n^{\beta,\omega}$ the path S is a Markov chain, with transition probabilities

$$P_n^{\beta,\omega}(S_{i+1}=y|S_i=x) = \frac{e^{\beta\omega(i+1,y)}Z_{n-i-1} \circ \theta_{i+1,y}}{Z_{n-i} \circ \theta_{i,x}} P(S_1=y|S_0=x) \tag{2.4}$$

for $0 \leq i < n$, and $P_n^{\beta,\omega}(S_{i+1} = y | S_i = x) = P(S_1 = y | S_0 = x)$ for $i \geq n$. Indeed, one directly sees that, for all path $(x_0 = 0, x_1, \ldots, x_n)$, the following product is telescopic,

$$\prod_{i=0}^{n-1} \frac{e^{\beta\omega(i+1,x_{i+1})} Z_{n-i-1} \circ \theta_{i+1,x_{i+1}}}{Z_{n-i} \circ \theta_{i,x_i}} P(S_1 = x_{i+1} | S_0 = x_i) = P_n^{\beta,\omega}(S_{[1,n]} = x_{[1,n]}),$$

which proves our claim. (Recall that $Z_0 = 1$.) We can re-write (2.4) as

$$P_n^{\beta,\omega}(S_{i+1} = y | S_i = x) = P_{n-i}^{\beta,\theta_{i,x}\omega}(S_1 = y - x)$$

The transition probabilities depend on the environment at the current time and at later times. Since the transition probability at step i depends on i, the chain is *time-inhomogeneous*. Since the transition probabilities also depend on the time horizon n, the family $P_n^{\beta,\omega}$ is not consistent: Except for the trivial case $\beta = 0$, there exists no Markov chain on the set of paths with infinite length which, for all $n \geq 1$, its marginal on the set of paths of length n is $P_n^{\beta,\omega}$. Therefore we consider the full sequence $(P_n^{\beta,\omega}; n \geq 1)$ of Markov chains. Finally, note that for $0 \leq m, n$,

$$P_{m+n}^{\beta,\omega}(S_{[1,n]} = \cdot | S_n = y) = P_n^{\beta,\omega}(S_{[1,n]} = \cdot | S_n = y),\tag{2.5}$$

$$P_{m+n}^{\beta,\omega}(S_{[n,n+m]} = y + \cdot | S_n = y) = P_m^{\beta,\theta_{n,y}\omega}(S_{[0,m]} = \cdot).\tag{2.6}$$

2.2 Free Energy

As it is well known in statistical mechanics, a great amount of information on the Gibbs measure is encoded in the following quantity,

$$p_n = p_n(\omega; \beta) = \frac{1}{n} \ln Z_n(\omega; \beta),\tag{2.7}$$

that we will call the (finite volume, specific) free energy[1] for the polymer of length n. The first step to understand its limit as the polymer length tends to infinity, and if it depends on the realization of the environment ω.

---

[1] In physics, the free energy is rather defined as $-\beta^{-1}p_n$, it has the same unit as the energy $-H_n$. The name specific means that it has been normalized by the number n of monomers.

Theorem 2.1 *As $n \to \infty$,*

$$p_n(\omega; \beta) \longrightarrow p(\beta) = \sup_n \frac{1}{n} \mathbb{P}\big[\ln Z_n(\omega; \beta)\big]$$

$\mathbb{P}$-*a.s. and in* L^p-*norm, for all* $p \in [1, \infty)$.

The theorem states that the sequence $p_n(\omega; \beta)$ converges a.s. to a limit, the limit is deterministic and given as a supremum over the polymer length. The limit p is called the (infinite volume, specific) free energy.[2]

Proof The proof splits in two steps, with the first one showing that expectations converge, and the second one showing that random fluctuations are negligible.

- *Step 1:* We first consider expected values and show that:

$$\lim_{n \to \infty} \mathbb{P}[p_n] = \sup_{n \in \mathbb{N}} \mathbb{P}[p_n] \in \mathbb{R}$$

For $m, n \geq 1$, recall the identity (2.3), and also that Z_m and $Z_m \circ \theta_{n,x}$ have the same law. By Jensen's inequality, we obtain

$$\ln Z_{n+m} \geq \ln Z_n + \sum_x P_n^{\beta,\omega}\{S_n = x\} \ln Z_m(\theta_{n,x}\omega) \ .$$

Taking expectation and using independence of the $\omega(i, y)$'s, we obtain

$$\mathbb{P}[\ln Z_{n+m}] \geq \mathbb{P}[\ln Z_n] + \sum_x \mathbb{P}[P_n^{\beta,\omega}\{S_n = x\}] \times \mathbb{P}[\ln Z_m]$$

$$= \mathbb{P}[\ln Z_n] + \mathbb{P}[\ln Z_m] \sum_x \mathbb{P}[P_n^{\beta,\omega}\{S_n = x\}]$$

$$= \mathbb{P}[\ln Z_n] + \mathbb{P}[\ln Z_m] \tag{2.8}$$

i.e., $\mathbb{P}[\ln Z_n]$ is super-additive. From the superadditive Lemma (see Toolbox, Sect. A.1), we see that

$$\lim_{n \nearrow \infty} \frac{1}{n} \mathbb{P}[\ln Z_n] = \sup_n \frac{1}{n} \mathbb{P}[\ln Z_n].$$

Now, the finiteness of p follows from the annealed bound (2.13) below.

---

[2]A much more detailed account will be given in Theorem 9.1 in the last chapter, defining the free energy in fixed directions.

- *Step 2:* We will apply to $\ln Z_n$ a concentration inequality, given in Theorem 2.2 below. Then, by Borel-Cantelli lemma, this implies that $\limsup_n |p_n - \mathbb{P}[p_n]| \leq \epsilon$, $\mathbb{P}$-a.s. Hence,

$$\limsup_{n \to \infty} |p_n - \mathbb{P}[p_n]| = 0 \qquad \mathbb{P} - \text{a.s.},$$

which, together with Step 1 above, completes the proof of almost sure convergence in Theorem 2.1. To get L^p convergence, one checks from the concentration inequality that the sequence $(p_n(\omega; \beta))_n$ is uniformly integrable. For definiteness, we will use the concentration inequality (2.9) in the second line below: since $EZ = \int_0^\infty P(Z > r)dr$ for $Z \geq 0$,

$$
\begin{aligned}
\mathbb{P}\big(|p_n - \mathbb{P}[p_n]|^p\big) &= \int_0^\infty \mathbb{P}\big(|p_n - \mathbb{P}[p_n]| > r^{1/p}\big)dr \\
&\leq 2\int_0^\infty \exp\{-nC(r^{1/p} \wedge r^{2/p})\}dr \\
&\leq 2\int_0^\infty \exp\{-nCr^{2/p}\}dr + 2\int_1^\infty \exp\{-nCr^{1/p}\}dr \\
&= C'n^{-p/2} + \mathcal{O}(\exp\{-Cn\})
\end{aligned}
$$

with $C' = \int_0^\infty \exp\{-Cv^{2/p}\}dv < \infty$. This implies L^p-convergence. $\qquad \square$

We have used a result of Liu and Watbled [167].

Theorem 2.2 (Theorem 1.4 in [167]) *(General concentration inequality for the free energy) Assume that the environment has all exponential moments (1.1). Then,*

$$\mathbb{P}\big[|p_n - \mathbb{P}[p_n]| \geq r\big] \leq \begin{cases} 2\exp\{-nCr^2\} & \text{if } 0 \leq r \leq 1, \\ 2\exp\{-nCr\} & \text{if } r \geq 1, \end{cases} \tag{2.9}$$

for some constant $C > 0$.

We will prove it here, only in two particular but important cases in Sect. A.3.1— Gaussian environment and bounded ones—, where the tails are subGaussian all the way (i.e., the bound $2\exp\{-nCr^2\}$ holds for all $r > 0$) and where the result follows from general principles. For bounded environment, it is a consequence of Azuma's Lemma A.2 as explained in Exercise A.1. For Gaussian environment the result follows from the Gaussian concentration inequality (A.4), and it takes the simplest possible form.

Remark 2.1 (Gaussian Environments) For a standard gaussian environment $\omega(t, x) \sim \mathcal{N}(0, 1)$, we can apply the well-known Gaussian concentration inequality (A.4), which yields

$$\mathbb{P}\big[|p_n - \mathbb{P}[p_n]| \geq \varepsilon\big] \leq 2 \exp\{-n\varepsilon^2/2\beta^2\}, \qquad \forall \varepsilon > 0. \tag{2.10}$$

Remark 2.2 (Self-averaging) The property that the random sequence $p_{n,\beta}^{\omega}$ becomes nonrandom in the limit is called self-averaging. Here, $p_{n,\beta}^{\omega}$ is the sum of $\mathbb{P}p_{n,\beta}^{\omega} = \mathcal{O}(n)$ and of a term of lower order $n^{-1/2}$ in probability according to the concentration inequality.

2.3 Upper Bounds

Computing the value of the free energy is impossible in general, hence we need to estimate it. We focus on upper bounds, and refer to Remark 2.5 for lower bounds.

2.3.1 The Annealed Bound

For all path S,

$$\mathbb{P}[\exp\{\beta H_n(S)\}] = \exp n\lambda(\beta), \tag{2.11}$$

and by Fubini's theorem,

$$\mathbb{P}[Z_n] = \mathbb{P}[P \exp\{\beta H_n(S)\}] = \exp n\lambda(\beta).$$

By Jensen inequality,

$$\mathbb{P}[p_n(\omega, \beta)] = \frac{1}{n}\mathbb{P}[\ln Z_n] \leq \frac{1}{n} \ln \mathbb{P}[Z_n] = \lambda(\beta), \tag{2.12}$$

hence

$$p(\beta) \leq \lambda(\beta). \tag{2.13}$$

This bound, comparing the quenched free energy p to the annealed free energy λ is central in the realm of random medium, it is known as the **annealed bound**. It appears in all standard disordered systems.

It is based on the single fact (2.11), and then averaging S under P. This bound is very crude, it does not depends either on: (1) the fine structure of P—except for being a probability measure—, (2) what are the correlations between $H_n(S)$ for different path S, (3) how rare are the deviations of $H_n(S)$ which are implicitly present in the value of $\mathbb{P}[\exp\{\beta H_n(S)\}]$.

In our case, the P-expectation performs an average on a collection of paths S which cardinality grows exponentially in n. Recall that Jensen's inequality is almost an equality when integrating a function which does not fluctuate much. Note also that correlations between the summands typically increase fluctuations of the average. Then, the annealed bound can be reasonable when the correlation between $H_n(S)$ for different S is small—e.g. if they are independent for different S's, cf. Lemmas 5.1, 5.3, 5.4 in [235]—and when the values of $H_n(S)$ contributing to that of $\mathbb{P}\exp\{\beta H_n(S)\}$ are not so rare in comparison with the entropy of P. For instance, for a stationary ergodic sequence $(\mathcal{E}_i)_{i\geq 1}$ with positive bounded values, we have as $n \to \infty$,

$$\mathbb{E}\ln\left(\frac{1}{n}\sum_{i=1}^{n}\mathcal{E}_i\right) \longrightarrow \ln\mathbb{E}(\mathcal{E})$$

by the ergodic theorem, and then the annealed bound is sharp in this asymptotics. In this example, we observe that the $(\mathcal{E}_i)_{i\geq 1}$ have a fixed correlation (which therefore should be viewed as weak as n diverges), and that the contributing values of $\mathcal{E}$ to the expectation are typical.

On the other hand, the annealed bound will be crude when the $H_n(S)$'s have strong correlation, or when the values of $H_n(S)$ contributing to that of $\mathbb{P}\exp\{\beta H_n(S)\}$ are too rare to be compensated by the entropy of P. To illustrate the first effect, consider the extreme case when the $H_n(S)$ are equal for all S to a nondegenerate random variable: Then, $\mathbb{E}\ln\left(\frac{1}{n}\sum_{i=1}^{n}\mathcal{E}_i\right) = \mathbb{E}\ln\mathcal{E}$ does not depend on n and is strictly less than $\ln\mathbb{E}(\mathcal{E})$ by Jensen inequality. In our framework, the annealed bound will not be sharp in general, due to these two effects. The first effect is more difficult to control than the second one, that we address in the next section.

Remark 2.3 (Annealed Bound Is Not Always Sharp) The inequality in annealed bound (2.13) may be strict. In this case of a Gaussian $\mathcal{N}(0, 1)$ distribution for ω, we have $\lambda(\beta) = \beta^2/2$, and for all path s, $H_n(\mathbf{x}) \sim \mathcal{N}(0, n)$. Recall the standard Gaussian tail estimate: if X has density $g(x) = (2\pi)^{-1/2}\exp\{-x^2/2\}$,

$$\frac{x}{1+x^2}g(x) \leq \mathbb{P}(X \geq x) \leq (\frac{1}{x} \wedge \sqrt{2\pi})g(x), \quad x > 0,$$

—see (A.2)—. Then, for all $a > 0$, we have by the union bound,

$$\sum_{n\geq 1} \mathbb{P}\left(\max_s H_n(s) > na\right) \leq \sum_{n\geq 1} \sum_{s:\text{length}\,n} \mathbb{P}\left(H_n(s) > na\right)$$

$$\leq \sum_{n\geq 1} (2d)^n \frac{1}{a\sqrt{n}} \exp\{-na^2/2\}$$

$$< \infty \qquad \text{if } a > \sqrt{2\ln(2d)}.$$

By Borel-Cantelli's lemma, we see that a.s.,

$$n^{-1} \ln Z_n \leq n^{-1}\beta \max_s\{H_n(s)\} \leq \beta a$$

for n large enough and such a's. Hence,

$$p(\beta) \leq \beta\sqrt{2\ln(2d)}\,. \tag{2.14}$$

Since this bound is of smaller order as $\beta \to \infty$ than $\lambda(\beta) = \beta^2/2$, we conclude that $p(\beta) < \lambda(\beta)$ for β large enough.

2.3.2 Improving the Annealed Bound

To improve (2.13) we will use monotonicity properties which are standard in thermodynamics:

Proposition 2.1 *For all fixed ω, we have:*

(i) The function $p_n(\omega;\beta)$ is convex in β, with $p_n(\omega;0) = 0$.
(ii) $\beta \mapsto \beta^{-1} p_n(\omega;\beta)$ is increasing.
(iii) $\beta \mapsto \beta^{-1}[p_n(\omega;\beta) + \ln(2d)]$ is decreasing.

The function $p(\beta)$ also satisfies (i)–(iii).

Proof Note that $\beta \mapsto p_n(\omega;\beta)$ is C^∞. By differentiation, one gets

$$\frac{d}{d\beta} n p_n = P_n^{\beta,\omega}[H_n]\,, \qquad \frac{d^2}{d\beta^2} n p_n = \mathrm{Var}_{P_n^{\beta,\omega}}[H_n] > 0\,,$$

proving (i). By convexity, $\beta^{-1} p_n(\beta,\omega) = \beta^{-1}[p_n(\beta,\omega) - p_n(0,\omega)]$ is non-decreasing in β. Turning to (iii), we have the identity

$$\frac{d}{d\beta}\left(\frac{1}{\beta}\left[p_n + \ln(2d)\right]\right) = -\frac{1}{\beta^2}\left[p_n + \ln(2d)\right] + \frac{1}{n\beta} P_n^{\beta,\omega}[H_n]$$

$$= \frac{1}{n\beta^2} h(P_n^{\beta,\omega})\,,$$

where $h(\nu)$ is the Boltzmann entropy of a probability measure ν on the n steps path space,

$$h(\nu) := \sum_{\mathbf{x}} \nu(\mathbf{x}) \ln \nu(\mathbf{x}) . \tag{2.15}$$

Clearly, $h(\nu) \leq 0$ for all ν, which ends the proof. $\qquad\square$

We derive an upper bound, which is improving on the annealed bound (2.13). It yields a sufficient condition for the strict inequality to hold in the annealed bound. It improves on the argument in the Remark 2.3, being however less transparent than the argument used in Remark 2.3. It is essentially of the same nature, in the sense that it relies on entropy considerations on a finite state space, and does not take into account the correlation structure of the random vector $(H_n(s))_s$.

Proposition 2.2 *We have*

$$p(\beta) \leq \beta \inf_{b \in (0,\beta]} \frac{\lambda(b) + \ln(2d)}{b} - \ln(2d) . \tag{2.16}$$

Hence, under Condition (T),

$$(\mathbf{T}): \qquad \beta\lambda'(\beta) - \lambda(\beta) > \ln(2d) , \tag{2.17}$$

we have

$$p(\beta) < \lambda(\beta).$$

More precisely, if there exists a positive root β_1 to the equation $\beta\lambda'(\beta) = \lambda(\beta) + \ln(2d)$, then for all $\beta > \beta_1$ it holds

$$p(\beta) \leq \frac{\beta}{\beta_1}\left[\lambda(\beta_1) + \ln(2d)\right] - \ln(2d) < \lambda(\beta) . \tag{2.18}$$

We summarize the above bounds in Figs. 2.1 and 2.2.

Proof To simplify the notation, we introduce

$$g(\beta) = \beta\lambda'(\beta) - \lambda(\beta) , \qquad f(\beta) = \frac{\lambda(\beta) + \ln(2d)}{\beta} . \tag{2.19}$$

Since λ is a smooth and convex, $g'(\beta) = \beta\lambda''(\beta)$ has the sign of β, and $g(\beta)$ is increasing on $\mathbb{R}_+$. In fact, g can be expressed from the convex conjugate λ^* of λ, with $\lambda^*(u) = \sup\{\beta u - \lambda(\beta)\}$ for $u \in \mathbb{R}$. We have

$$g(\beta) = \lambda^*(u) , \qquad u = \lambda'(\beta).$$

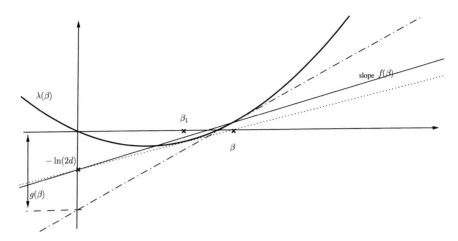

Fig. 2.1 Constructing the upper bound of Proposition 2.2; $f(\beta)$ is the slope of the *solid line* through the points $(0, -\ln(2d))$ and $(\beta, \lambda(\beta))$; $g(\beta)$ is the y-intercept of the tangent to λ at β. For β as in the figure, the infimum of $f(\beta') = \frac{\lambda(\beta') + \ln(2d)}{\beta'}$ over the interval $[0, \beta]$ is achieved at $\beta' = \beta_1$ where the tangent to λ intersects the vertical axis at $-\ln(2d)$

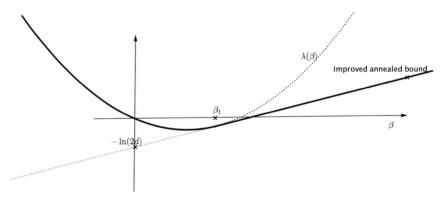

Fig. 2.2 Improved annealed bound. If λ has a tangent at some $\beta_1 > 0$ which intersects the vertical axis at height $-\ln(2d)$, the bound for p on $\mathbb{R}^+$ is improved on the annealed bound $\lambda(\beta)$ by that tangent for $\beta > \beta_1$

Moreover, $f'(\beta) = (g(\beta) - \ln(2d))/\beta^2$. We can write

$$\mathbb{P}[p_n(\omega, \beta) + \ln(2d)] \quad = \quad \beta \times \frac{1}{\beta} \mathbb{P}[p_n(\omega, \beta) + \ln(2d)]$$

$$\overset{\text{Proposition 2.1 (iii)}}{=} \quad \beta \times \inf_{b \in (0,\beta]} \frac{1}{b} \mathbb{P}[p_n(\omega, b) + \ln(2d)]$$

$$\overset{(2.12)}{\leq} \quad \beta \inf_{b \in (0,\beta]} f(b) \,.$$

Letting $n \to \infty$, we obtain

$$p(\beta) + \ln(2d) \leq \beta \inf_{b \in (0, \beta]} \frac{\lambda(b) + \ln(2d)}{b}, \tag{2.20}$$

which is the bound (2.16). Now, let us define $\beta_1 \in (0, \infty]$ by

$$\beta_1 = \inf\{\beta \geq 0 : g(\beta) \geq \ln(2d)\},$$

allowing an infinite value. Then, f has its minimum at β_1. Then, as shown in Fig. 2.1,

$$\inf_{\beta' \in]0, \beta]} f(\beta') = \begin{cases} f(\beta) & \text{if } \beta \leq \beta_1, \\ f(\beta_1) & \text{if } \beta \geq \beta_1. \end{cases}$$

With (2.20), this yields the bound in (2.18). Note that $\beta > \beta_1$ is equivalent to Condition (**T**). For such a β, the last inequality is strict by strict convexity of λ. $\quad\square$

Remark 2.4

(i) We could have used the notation $\beta_1 = \beta_c^{\mathrm{tr}}$, since it is equal to the critical value for the polymer model on the tree, see Theorem 4.2. In fact the bound given in (2.16) is the free energy for the model on the tree.

(ii) The improved upper bound follows from Proposition 2.1 (iii) and an explicit computation for the simple random walk. Similar improvement can be implemented for general a priori laws P of the path, as long-range walks. The only requirement is an upper bound on the entropy rate.

Example 2.1 We consider again the Gaussian case, $\omega(i, x) \sim \mathcal{N}(0, 1)$. We easily compute $\beta_1 = \sqrt{2 \ln(2d)}$, we check that the bound in (2.18) is equal to $\beta \sqrt{2 \ln(2d)} - \ln(2d)$, and therefore is strictly smaller than the one from (2.14).

We now look for conditions ensuring $p(\beta) < \lambda(\beta)$ for large β, in terms of the marginal distribution q of ω. A sufficient condition is that the environment is unbounded from above, or bounded from above with a large enough mass on its essential supremum.

Corollary 2.1 *Set $q(dh) = \mathbb{P}(\omega(n, x) \in dh)$ and $s = \sup \operatorname{supp}[q]$. If $s = +\infty$ or $q(\{s\}) < \frac{1}{2d}$, then, there exists $\beta_1 \in (0, \infty)$ such that $p(\beta) < \lambda(\beta)$ for $\beta > \beta_1$.*

Proof We check condition (2.17). Let λ^* the Legendre transform of λ,

$$\lambda^*(u) = \sup\{u\beta - \lambda(\beta)\}$$

Then,

$$\beta\lambda'(\beta) - \lambda(\beta) = \lambda^*(\lambda'(\beta))$$
$$= \tilde{q}^\beta \Big[\ln \frac{d\tilde{q}^\beta}{dq} \Big]$$

where $\tilde{q}^\beta$ is the tilted probability measure on $\mathbb{R}$ given by

$$\tilde{q}^\beta(d\omega) = e^{\beta\omega - \lambda(\beta)} q(d\omega)$$

As illustrated in Fig. 2.1, $\lambda'(\beta) \to \sup \operatorname{supp} q$ when $\beta \to +\infty$.

1. If $s = +\infty$, then $\lambda^* \to \infty$ at infinity, and (2.17) holds for large β.
2. If $s < +\infty$,

$$\lim_{\beta \to +\infty} \lambda^*(\lambda'(\beta)) = -\ln q(\{s\})$$

and (2.17) holds for large β if we assume $q(\{s\}) < 1/2d$. □

Remark 2.5 Lower bounds are less useful for our analysis. We can use the formula as a supremum from Theorem 2.1, and the simplest application leads to

$$p(\beta) \geq \mathbb{P}[p_1(\beta)] = \mathbb{P} \ln P[\exp\{\beta\omega(1, S_1)\}]$$

which is already better than using Jensen inequality

$$p(\beta) \geq \ln P[\exp\{\beta\mathbb{P}[\omega(1, S_1)]\}] = \beta\mathbb{P}[\omega(t, x)].$$

But all these bounds correspond to local optimization in contrast with the polymer measure which performs optimization in highly non local manner.

2.4 Monotonicity

We have seen in Proposition 2.1 that $\beta \mapsto p_n(\omega; \beta)$ is convex and differentiable. It is plain to compute its derivative,

$$\frac{\partial}{\partial\beta} p_n(\omega; \beta) = \frac{1}{n} \frac{\partial}{\partial\beta} \ln Z_n(\omega; \beta) = P_n^{\beta, \omega}[H_n(S)/n],$$

which is the specific (internal) energy. Let $\mathscr{D} = \mathscr{D}(p)$ be the set of β's such that the limit p is differentiable at β. Since p is convex, it is a general fact from convex functions theory that its complement $\mathscr{D}^c$ is at most countable. In fact, p is $\mathscr{C}^1$ on $\mathscr{D}$. Since p is the limit of $p_n(\omega; \beta)$ a.s. and in the L^1-norm, we deal with a third convex function, e.g., $\mathbb{P}[p_n(\omega; \beta)]$. The following is a direct consequence of general results on convergence of convex functions and their derivatives.

Proposition 2.3 *For all $\beta \in \mathscr{D}$ and almost every environment ω,*

$$\lim_{n \to \infty} P_n^{\beta,\omega}[H_n(S)/n] = \lim_{n \to \infty} \mathbb{P}[P_n^{\beta,\omega}(H_n(S)/n)] = p'(\beta).$$

Moreover, for all $\beta \in \mathbb{R}$,

$$p'(\beta^-) \leq \liminf_{n \to \infty} P_n^{\beta,\omega}[H_n(S)/n] \leq \limsup_{n \to \infty} P_n^{\beta,\omega}[H_n(S)/n] \leq p'(\beta^+).$$

The notation $p'(\beta^+)$ denotes the limit of $p'(b)$ as b decreases to β in $\mathscr{D}$. We could as well write bounds involving the left and right derivatives. In this section, we show that

$$\mathbb{P}[\omega(t, x)] \leq p'(\beta) \leq \lambda'(\beta), \qquad \beta \geq 0, \beta \in \mathscr{D},$$

where the last inequality follows from (2.22) below. But the main point is that the difference $\lambda - p$ has a nice monotonicity property.

Theorem 2.3 *The functions $\beta \mapsto \lambda(\beta) - \mathbb{P}[p_n(\omega; \beta)]$ and $\beta \mapsto \lambda(\beta) - p(\beta)$ are non-decreasing on $\mathbb{R}^+$, and non-increasing on $\mathbb{R}^-$.*

Proof Recall the notation q for the law of $\omega(t, x)$ under $\mathbb{P}$. With

$$\zeta_n(S) = \exp \beta H_n(S),$$

it is straightforward to check

$$\frac{\partial}{\partial \beta} \mathbb{P}[\ln Z_n] = \mathbb{P}\left[\frac{\partial}{\partial \beta} \ln Z_n\right]$$

$$= \mathbb{P}[(Z_n)^{-1} \frac{\partial}{\partial \beta} Z_n]$$

$$= P\left[\mathbb{P}[(Z_n)^{-1} H_n \zeta_n]\right]$$

At this point, we will use the fact that independent variables are positively associated, and satisfy the Harris-FKG inequality given in the toolbox.

Fix a path $\mathbf{x}$. Then, the probability measure $\tilde{\mathbb{P}}^{\mathbf{x}}$,

$$d\tilde{\mathbb{P}}^{\mathbf{x}} = \zeta_n(\mathbf{x})e^{-n\lambda(\beta)}d\mathbb{P}$$

is product, and therefore the family ω satisfies the FKG inequality. Indeed, the variables $\omega(t,x), t \geq 0, x \in \mathbb{Z}^d$, are independent under $\tilde{\mathbb{P}}^{\mathbf{x}}$—though not identically distributed—with $\omega(t,x)$ distributed as q or $d\tilde{q}^{\beta}(w) = e^{\beta w - \lambda}dq(w)$ according to $x_t \neq x$ or $x_t = x$. Note that the function H_n is increasing in ω, while $(Z_n)^{-1}$ is a decreasing for $\beta \geq 0$. We apply Proposition A.1 for fixed $\mathbf{x}$, and we find

$$\mathbb{P}[(Z_n)^{-1}H_n\zeta_n] \leq e^{-n\lambda(\beta)}\mathbb{P}[(Z_n)^{-1}\zeta_n] \times \mathbb{P}[H_n\zeta_n]$$
$$= \mathbb{P}[(Z_n)^{-1}\zeta_n] \times n\lambda'(\beta)$$

using independence and also that

$$\mathbb{P}[\omega(t,x)e^{\beta\omega(t,x)}] = \lambda'(\beta)e^{\lambda(\beta)} . \tag{2.21}$$

Integrating with respect to P, we get

$$\frac{\partial}{\partial\beta}P\ln Z_n \leq n\lambda'(\beta)P\left[\mathbb{P}[(Z_n)^{-1}\zeta_n]\right]$$
$$= n\lambda'(\beta)\mathbb{P}\left[(Z_n)^{-1}P[\zeta_n]\right]$$
$$= n\lambda'(\beta) \tag{2.22}$$

which yields the desired result, since $p_n(\omega;\beta)$ and λ are both equal to zero when $\beta = 0$. $\qquad\square$

For more information on FKG inequality, see [165, pp. 77–83].

Remark 2.6 It is proved in [188] that the r.v.'s $Z_n(\beta,\omega)e^{-n\lambda(\beta)}$ are increasing in β in the convex order. For two integrable r.v.'s X, X', we say that X is smaller than X' in convex order if

$$\mathbb{P}[\phi(X)] \leq \mathbb{P}[\phi(X')],$$

for all convex $\phi : \mathbb{R} \to \mathbb{R}$ such that the expectations exist. We write $X \leq_{cx} X'$. First, this implies that

$$\mathbb{P}[X] = \mathbb{P}[X'],$$

since both functions $\phi(x) = x$ and $\phi(x) = -x$ are convex, and furthermore,

$$\text{Var}(X) \leq \text{Var}(X'), \quad \mathbb{P}[|X - a|^q] \leq \mathbb{P}[|X' - a|^q] \quad (a \in \mathbb{R}, q \geq 1).$$

Since they have same expectation, this is a natural quantitative way to express that X is less dispersed than X'.

In our case, the process $\beta \mapsto Z_n(\beta, \omega)e^{-n\lambda(\beta)}$ is increasing for the convex order. It is a mathematical formulation of the intuitive property that

the fluctuations of $Z_n(\beta, \omega)e^{-n\lambda(\beta)}$ increase as β grows.

An increasing process for the convex order is called a peacock. It is well known that it has the same marginals as some martingale: There exists a martingale $(M(\beta), \beta \in \mathbb{R}_+)$ such that for all β,

$$Z_n(\beta, \omega)e^{-n\lambda(\beta)} \overset{\text{law}}{=} M(\beta).$$

(Of course, M depends also on n.) Such a martingale M is explicitly known in a few examples (Gaussian or Bernoulli environment), but not in general. See [188] for more details.

2.5 Phase Transition

In this section, we elaborate a phase diagram based on whether the annealed bound (2.13) is achieved or not.

Theorem 2.4 (Critical Temperature) *There exists* $\beta_c = \beta_c(\mathbb{P}, d) \in [0, \infty]$ *such that*

$$\begin{cases} p(\beta) = \lambda(\beta) & \text{if } 0 \leq \beta \leq \beta_c, \\ p(\beta) < \lambda(\beta) & \text{if } \beta > \beta_c \end{cases} \tag{2.23}$$

Proof Define

$$\beta_c = \inf\{\beta \geq 0 : p(\beta) < \lambda(\beta)\}. \tag{2.24}$$

The claim is a direct consequence of Proposition 2.3. □

The terminology in the next definition becomes transparent.

Definition 2.1 We call *high temperature region* (or small β region) the set of β's such that $p = \lambda$, and the *low temperature region* (or large β region) the set of β's such that $p < \lambda$.

We can already make a few observations:

1. We have trivially $p(0) = 0 = \lambda(0)$, showing that $\beta = 0$ is in the high temperature region.
2. We have seen sufficient conditions for $\beta_c < \infty$, e.g., condition (T) in (2.17).

3. Theorem 2.4 implies the absence of reentrant phase transition in the phase
   diagram of the model. Of course, in complete generality, we may have $\beta_c = 0$
   or ∞, i.e., absence of one of the two regimes in the interval $(0, \infty)$.

One expects that the polymer measure has completely different behavior in
these two regions. In the high temperature region, the Gibbs measure is a small
perturbation of the simple random walk. At low temperature, the polymer strongly
feels the environment, and should have quite different scaling limits.

Breaking the Analyticity Recall that a function is analytic at a point β_0 in the
interior of its domain if it is equal in a neighborhood of β_0 to the sum of a power
series in $\beta - \beta_0$ with a positive radius of convergence. For instance, λ is analytic
in $\mathbb{R}$. Recall that $\beta \mapsto p_n(\omega; \beta)$ is infinitely differentiable; since it is the logarithm
of a finite sum of smooth terms, it is easy to see that it is even analytic for all n.
However, the limit p may not be analytic.

If β_c is strictly positive and finite, both high and low temperature regions have a
nonempty interior, the function p is nonanalytic at $\beta = \beta_c$. (Indeed, it is given by
$p(\beta) = \lambda(\beta)$ for $\beta \in [0, \beta_c]$, which analytic continuation on $\mathbb{R}$ is $\lambda(\beta)$, under the
assumption (1.1).) The value β_c is called **critical**.

Physical Picture In physics, a phase transition is the passage of a thermodynamic
system from one phase to another. The distinguishing characteristic of a phase
transition is an abrupt change in one or more physical properties, in particular
the heat capacity, with a small change in a thermodynamic variable such as the
temperature. To keep our discussion simple, we take some freedom with statistical
mechanics, considering the energy H_n as only observable, and β as the only
parameter.

- The function p is analytic at a point β_0 if it is the sum of a power series in $\beta - \beta_0$
  with a positive radius of convergence.
- If p is non-analytic at β_0, β_0 is called a critical point.
- Ehrenfest proposed a classification scheme, grouping phase transitions based
  on the degree of non-analyticity: phase transitions are labelled by the lowest
  derivative of the free energy that is discontinuous at the transition. First-order
  phase transitions exhibit a discontinuity in the first derivative of the free energy
  (with respect to some thermodynamic variable).

Important questions are:

- What phenomenon is responsible for the phase transition? In what respect the
  polymer measures below and above the critical value are different? If we observe
  a sample of paths from the Gibbs measure, can we decide if $\beta > \beta_c$ or not?
- What happens at the critical value? Critical phenomena are believed to depend on
  very few details of the model, which are then grouped into universality classes.
  Loosely speaking, an universality class is the collection of all models sharing the
  same scalings and/or the same limits at criticality.

We anticipate the next chapters and mention at this point that high temperature corresponds to the delocalized phase, and low temperature to the localized phase. We conclude this section with a simple phase diagram, depicting phases according to the parameter value β, that we restrict to $[0, +\infty)$ for the case of negative β is covered by changing the environment ω into $-\omega$:

β	0	β_c	$+\infty$
Temperature	High temperature	Low temperature	
Free energy	$p = \lambda$	$p < \lambda$	
Phase	Delocalized	Localized	

Chapter 3
The Martingale Approach and the L^2 Region

Martingale theory is well-known as a powerful tool to study random sequences. In this section, we start to use it in our context. First of all, it is efficient for proving that equality can hold in (2.13). The simplest sufficient condition for equality is based on a second moment computation. The parameter region where second moment computations are reliable, will be called the L^2 region.

3.1 A Martingale Associated to the Partition Function

Classical considerations from thermodynamics and common sense made us consider $\ln Z_n(\omega; \beta)$, a rather difficult quantity to study directly since we cannot even compute its expectation! It is far easier to consider the partition function $Z_n(\omega; \beta)$ itself, for which we easily see that $\mathbb{P}[Z_n(\omega; \beta)] = \exp\{n\lambda(\beta)\}$. All through, we will consider the *normalized partition function*, defined by

$$W_n = Z_n(\omega; \beta) \exp(-n\lambda(\beta)), \quad n \geq 1. \tag{3.1}$$

This variable has expectation 1, which explains why we call it "normalized". To keep notations simple, we drop the arguments β and ω.

Fix a path $\mathbf{x}$ of the simple random walk. Then, the sequence $(H_k(\mathbf{x}))_k$ is the sum of independent identically distributed real random variables on $(\Omega, \mathcal{G}, \mathbb{P})$—the randomness coming from the environment alone—, it is itself a random walk. The corresponding exponential martingale is

$$\bar{\zeta}_n = \bar{\zeta}_n(\mathbf{x}) = \exp\left(\beta H_n(\mathbf{x}) - n\lambda(\beta)\right), \tag{3.2}$$

for $\beta \in \mathbb{R}$: this is a positive, mean 1 martingale on $(\Omega, \mathcal{G}, \mathbb{P})$, with respect to the filtration $(\mathcal{G}_n)_n$, where

$$\mathcal{G}_n = \sigma\{\omega(j, x) ; j \leq n, \ x \in \mathbb{Z}^d\},$$

© Springer International Publishing AG 2017
F. Comets, *Directed Polymers in Random Environments*, Lecture Notes
in Mathematics 2175, DOI 10.1007/978-3-319-50487-2_3

This holds for all path $\mathbf{x}$. By making a linear combination of those martingales indexed by $\mathbf{x}$, we will get another martingale. In particular,

$$W_n = P(\bar{\zeta}_n) \quad \text{is a positive martingale.}$$

This is much stronger a property than $\mathbb{P}[W_n] = 1$, and it will make the sequence $Z_n(\omega; \beta)$ much easier to study than $\ln Z_n(\omega; \beta)$ itself, a fact which was first noticed by Bolthausen in his pioneering work [46].

By Doob's martingale convergence theorem [239, Corollary 11.7], the limit W_∞ exists $\mathbb{P}$-a.s., and is non-negative.

At this point one naturally wonders if the limit W_∞ is 0 or not. It is well-known that, for $\beta \neq 0$, the martingale $\bar{\zeta}_n$ vanishes for all infinite paths $\mathbf{x}$,

$$\bar{\zeta}_n \to 0 \quad \text{a.s. as } n \to \infty. \tag{3.3}$$

However, their average $W_n = P[\bar{\zeta}_n]$ may converge to a positive limit. This may sound paradoxical at first, but it can be understood as follows: Though the martingale $\bar{\zeta}_n$ attached to a single fixed path ω is doomed to vanish, its mean over many paths can survive for the diffusion effect allows collecting "big" contributions from distant parts of the space. The number of paths contributing to W_n is $(2d)^n$; being exponential in n, it has a chance to balance the decay of $\bar{\zeta}_n$ in (3.3), which itself decays exponentially in n by large deviation theory. If the diffusion dominates the random fluctuations of the environment, the vanishing of (3.3) will be compensated in W_n and the limit will be nonzero.

It is not difficult to see that the event $\{W_\infty = 0\}$ is measurable with respect to the tail σ-field

$$\mathscr{T} = \bigcap_{n \geq 1} \mathscr{T}_n, \quad \mathscr{T}_n = \sigma\{\omega(j, x) ; j \geq n, x \in \mathbb{Z}^d\}.$$

Indeed, from the Markov property (2.3) we can write

$$W_{n+m} = P\big[\bar{\zeta}_n \times W_m \circ \theta_{n, S_n}\big],$$

and then

$$\begin{aligned}
W_\infty &= \lim_{m \to \infty} W_{n+m} \\
&= P\big[\bar{\zeta}_n \times \lim_{m \to \infty} W_m \circ \theta_{n, S_n}\big] \quad \text{(finite sum)} \\
&= W_n \times \sum_{x \in \mathbb{Z}^d} P_n^{\beta, \omega}(S_n = x) W_\infty \circ \theta_{n, x}
\end{aligned}$$

For all n, by strict positivity of $\bar{\zeta}_n$, the event under consideration is equal to

$$\{W_\infty = 0\} = \bigcap_{x \in \mathbb{Z}^d : P(S_n=x)>0} \left\{ W_\infty \circ \theta_{n,x} = 0 \right\}.$$

We conclude that $\{W_\infty = 0\} \in \mathscr{T}_n$ for all n, and then $\{W_\infty = 0\} \in \mathscr{T}$. By Kolmogorov's zero-one law [104], every event in the tail σ-field $\mathscr{T}$ has probability 0 or 1. Summarizing all this, we can state the following.

Theorem 3.1 *The limit*

$$W_\infty = \lim_{n \nearrow \infty} W_n \tag{3.4}$$

exists $\mathbb{P}$-a.s. Moreover, we have a dichotomy. Either the limit W_∞ is a.s. positive, or it is a.s. zero:

$$\mathbb{P}\{W_\infty > 0\} = 1, \tag{3.5}$$

or

$$\mathbb{P}\{W_\infty = 0\} = 1. \tag{3.6}$$

We introduce some terminology for the parameter regions where one of these two contrasting situations hold.

Definition 3.1 The polymer is the **weak disorder** phase when (3.5) holds, and in the **strong disorder** phase when (3.6) holds.

Again, observe that, for $\beta = 0$, $W_n = 1$ for all n, so that weak disorder takes place. We will see later that the polymer is diffusive in the regime (3.5), as well as other consequences.

Remark 3.1

(i) It is an interesting question to find a characterization of (3.5) [or (3.6)] in terms of the distribution of $\omega(n, x)$. It will be addressed in Sect. 5.1 below.

(ii) It is not difficult to see that

$$W_\infty > 0 \implies p(\beta) = \lambda(\beta). \tag{3.7}$$

Indeed, we have a.s.

$$p(\beta) = \lim_{n \to \infty} n^{-1} \ln Z_n(\omega; \beta) = \lambda(\beta) + \lim_{n \to \infty} n^{-1} \ln W_n,$$

where $\lim_{n \to \infty} \ln W_n = \ln W_\infty$ is finite if $W_\infty > 0$. Further, with the convexity and differentiability of the limit, it implies $\lim_n n^{-1} P_n^{\beta,\omega}[H_n] = \lambda'(\beta)$. Even more, it implies that

$$P_n^{\beta,\omega}\left[\left|H_n/n - \lambda'(\beta)\right|\right] \to 0 \tag{3.8}$$

as $n \to \infty$, see e.g. [15, Proposition 3].

(iii) Except for $\beta = 0$, the random variable W_∞ is not $\mathscr{T}$-measurable in the weak disorder phase. To see this, observe that it has to depend in a significant manner of $\omega(1, x)$ with x nearest neighbor of the origin.

It is natural to introduce another critical value of the parameter.

Proposition 3.1 *There exists* $\bar{\beta}_c = \bar{\beta}_c(\mathbb{P}, d) \in [0, \infty]$ *such that*

$$\begin{cases} W_\infty > 0 & a.s. \quad if \ \beta \in \{0\} \cup (0, \bar{\beta}_c), \\ W_\infty = 0 & a.s. \qquad if \ \beta > \bar{\beta}_c \end{cases} \tag{3.9}$$

A number of remarks analogous to those in Sect. 2.5 can be formulated here, including:

1. β measures the strength of the disorder;
2. There exist sufficient conditions for $\bar{\beta}_c > 0$ (cf. condition (L2) in (2.1)) and for $\bar{\beta}_c < \infty$ (see (3.12) below);
3. The phase diagram weak versus strong disorder has no reentrant phase transition.

Proof Let $\delta \in (0, 1)$ arbitrary. Since $\mathbb{P}[W_n] = 1$, the sequence $(W_n^\delta)_n$ is uniformly integrable. In addition to a.s. convergence of W_n^δ to W_∞^δ this implies that

$$\lim_{n\to\infty} \mathbb{P}[W_n^\delta] = \mathbb{P}[W_\infty^\delta] \,,$$

which is either 0 in the strong disorder case, or strictly positive in the weak disorder case. We claim that

$$\beta \mapsto \mathbb{P}[W_n^\delta] \quad \text{is non-increasing on } \mathbb{R}_+ \,. \tag{3.10}$$

This will imply that $\mathbb{P}[W_\infty^\delta]$ also is non-increasing on the positive half-line. Then, we obtain the proposition by putting $\bar{\beta}_c = \inf\{\beta \geq 0 : \mathbb{P}[W_\infty^\delta] = 0\}$ (with the convention $\inf \emptyset = +\infty$). It only remains to prove (3.10). We have

$$\begin{aligned} \frac{d}{d\beta}\mathbb{P}[W_n^\delta] &= \mathbb{P}\left[\frac{d}{d\beta}W_n^\delta\right] \\ &= \delta\mathbb{P}\left[W_n^{\delta-1}P\{(H_n - n\lambda')\bar{\zeta}_n\}\right] \\ &= \delta P\left[\mathbb{P}\{\bar{\zeta}_n W_n^{\delta-1}(H_n - n\lambda')\}\right] \\ &\leq \delta P\left[\mathbb{P}\{\bar{\zeta}_n W_n^{\delta-1}\}\,\mathbb{P}\{\bar{\zeta}_n(H_n - n\lambda')\}\right] \qquad \text{(by FKG)} \\ &= 0 \qquad\qquad\qquad\qquad\qquad\qquad\qquad\qquad \text{(by 2.21)} \end{aligned}$$

The FKG inequality from Proposition A.1 was applied above to the product measure $\bar{\zeta}_n d\mathbb{P}$ for fixed $\mathbf{x}$, and to the decreasing function $W_n^{\delta-1}$ of ω and the non-decreasing one $H_n - n\lambda'$. The details being completely identical to those of the proof of Proposition 2.3, we omit them here. □

Open Problem 3.2 *Many basic questions are left open so far:*

(i) *Is $\beta_c = \bar{\beta}_c$? The inequality $\beta_c \geq \bar{\beta}_c$ holds trivially by (3.7). Does (3.6) in a neighborhood of β_0 implies $p(\beta) < \lambda(\beta)$ in a neighborhood of β_0? As we will see in Sect. 6.2, the answer is yes when $d = 1$ and $d = 2$ [157]. But this will come out by showing that $\beta_c = 0$. The question is open in dimension $d \geq 3$.*

(ii) *What happens at criticality? When $\bar{\beta}_c \in (0, \infty)$, is W_∞ zero or non zero for $\beta = \bar{\beta}_c$?*

For both questions, conjectures can be made on the basis of related models of branching processes, multiplicative cascades and percolation: the predictions are

Conjecture 3.1

$$\beta_c = \bar{\beta}_c, \qquad W_\infty(\bar{\beta}_c) = 0.$$

3.2 The Second Moment Method and the L^2 Region

In this section, we show how to prove that weak disorder holds for some values of the parameters d and β. The proof will be based on a second moment computation. The second moment of the normalized partition function can be written explicitly in terms of the expectation of a function of two independent copies of the random walk, the function being the exponent of the number of intersections between the walks.

We first recall the following fact about the return probability π_d for the simple random walk,

$$\pi_d \stackrel{\text{def.}}{=} P\{S_n = 0 \text{ for some } n \geq 1\} \quad \text{is} \quad \begin{cases} = 1 & \text{if } d \leq 2, \\ < 1 & \text{if } d \geq 3. \end{cases} \tag{3.11}$$

More precisely, it is known that $\pi_{d+1} < \pi_d$ for all $d \geq 3$ (e.g., [191, Lemma 1]) and that $\pi_3 = 0.3405\ldots$ [219, p. 103]. In particular, $\pi_d \leq 0.3405\ldots$ for all $d \geq 3$.

Theorem 3.3 ([46]) *Suppose that $d \geq 3$ (hence $\pi_d < 1$) and that condition (L2) holds:*

$$\textbf{(L2)} \qquad \lambda_2(\beta) \stackrel{\text{def.}}{=} \lambda(2\beta) - 2\lambda(\beta) < \ln(1/\pi_d). \tag{3.12}$$

Then, $W_\infty > 0$ a.s.

Note first that $\lambda_2(\beta)$ is continuous with $\lambda_2(0) = 0$, so that, for $d \geq 3$, the condition (L2) does hold if β is small, whatever the distribution of the environment is. More precisely, since λ is increasing, $\lambda_2'(\beta) = 2[\lambda'(2\beta) - \lambda'(\beta)]$ is positive for $\beta \geq 0$, and so $\lambda_2(\beta)$ is increasing on $\mathbb{R}^+$; similarly, it is decreasing on $\mathbb{R}^-$. For instance, in the case of a standard Gaussian distribution for ω, we have $\lambda_2(\beta) = \beta^2$, and when ω is Bernoulli variable taking values ± 1 with probability $1/2$, $\lambda_2(\beta) = \ln(1 + \tanh^2(\beta))$.

The point is that, for $d \geq 3$, the condition (L2) is equivalent to $\beta < \beta_{L^2}$ with

$$\beta_{L^2} = \inf\{\beta \geq 0 : \lambda_2(\beta) \leq \ln(1/\pi_d)\}. \tag{3.13}$$

In particular, $p = \lambda$ holds for $\beta \leq \beta_{L^2}$.

Definition 3.2 The set of β's defined by the condition (L2), i.e. (3.12), is called the L^2 region. In dimension $d \geq 3$, it consists in a non-empty interval $(0, \beta_{L^2})$, which is a subset of the weak disorder region, and thus, of the high temperature region.

Proof (Proof of Theorem 3.3) We compute the L^2-norm of the martingale W_n. To do so, we consider on the product space $(\Omega^2, \mathscr{F}^{\otimes 2})$, the probability measure $P^{\otimes 2} = P^{\otimes 2}(d\mathbf{x}, d\widetilde{\mathbf{x}})$, that we will view as the distribution of the couple $(S, \widetilde{S})$ with $\widetilde{S} = (\widetilde{S}_k)_{k \geq 0}$ an independent copy of $S = (S_k)_{k \geq 0}$. An elementary, though important, observation is that

$$W_n^2 = P^{\otimes 2}\left[e^{\beta[H_n(S) + H_n(\widetilde{S})] - 2n\lambda(\beta)}\right].$$

Hence, by Fubini and by independence,

$$\mathbb{P}[W_n^2] = P^{\otimes 2}\mathbb{P}\left[\prod_{t=1}^n e^{\beta[\omega(t,S_t) + \omega(t,\widetilde{S}_t)] - 2\lambda(\beta)}\right]$$

$$= P^{\otimes 2}\left[\prod_{t=1}^n \mathbb{P}\left(e^{\beta[\omega(t,S_t) + \omega(t,\widetilde{S}_t)] - 2\lambda(\beta)}\left(\mathbf{1}_{S_t = \widetilde{S}_t} + \mathbf{1}_{S_t \neq \widetilde{S}_t}\right)\right)\right]$$

$$= P^{\otimes 2}\left[\prod_{t=1}^n \left(e^{\lambda(2\beta) - 2\lambda(\beta)}\mathbf{1}_{S_t = \widetilde{S}_t} + \mathbf{1}_{S_t \neq \widetilde{S}_t}\right)\right]$$

$$= P^{\otimes 2}\left[\prod_{t=1}^n e^{\lambda_2(\beta)\mathbf{1}_{S_t = \widetilde{S}_t}}\right]$$

$$= P^{\otimes 2}\left[e^{\lambda_2(\beta)N_n}\right],$$

with N_n the number of intersections of the paths $S, \widetilde{S}$ up to time n,

$$N_n = N_n(S, \widetilde{S}) = \sum_{t=1}^n \mathbf{1}_{S_t = \widetilde{S}_t} \tag{3.14}$$

As $n \to \infty$, $N_n \nearrow N_\infty$, and by monotone convergence $\mathbb{P}[W_n^2] \nearrow P^{\otimes 2}\left[e^{\lambda_2(\beta)N_\infty}\right]$. Observe that

$$N_n = \sum_{t=1}^{2n} 1_{\hat{S}_t=0},$$

where

$$\hat{S}_0 = 0, \ \hat{S}_1 = S_1, \ \hat{S}_2 = S_1 - \tilde{S}_1, \ \hat{S}_3 = S_1 + S_2 - \tilde{S}_1, \ldots \text{ for } k \geq 0,$$

$$\hat{S}_{2k+1} = \hat{S}_{2k} + S_{k+1}, \ \hat{S}_{2k+2} = \hat{S}_{2k+1} - \tilde{S}_{k+1},$$

is itself a symmetric simple random walk on $\mathbb{Z}^d$. Therefore, the number N_∞ of visit to 0 of this new simple random walk, is geometrically distributed with success probability π_d. Finally, as $n \to \infty$,

$$\mathbb{P}[W_n^2] \nearrow P^{\otimes 2}\left[e^{\lambda_2(\beta)N_\infty}\right] = \sum_{k=0}^{\infty}(1-\pi_d)\pi_d^k e^{k\lambda_2}$$

$$= \begin{cases} \frac{1-\pi_d}{1-\pi_d e^{\lambda_2}} & \text{if } \pi_d e^{\lambda_2} < 1 \\ +\infty & \text{if } \pi_d e^{\lambda_2} \geq 1 \end{cases}$$

Therefore,

$$\sup_n \mathbb{P}[W_n^2] < \infty \iff \lambda_2 + \ln \pi_d < 0 \,,$$

i.e., iff (3.12) is fulfilled. Then, the martingale W_n is bounded in L^2, and by a classical convergence result [239], it converges in L^2 to a limit, which is necessarily equal to W_∞. Therefore, we have convergence in L^1, and $\mathbb{P}[W_\infty] = \lim_n \mathbb{P}[W_n] = 1$, which excludes the possibility that the limit vanishes in Theorem 3.1.   □

Remark 3.2 Finer sufficient conditions for weak disorder, improving on (3.12), have been recently obtained: [38] making use of size-biasing (we will present this technique in section 5.3); [55] by a comparison of environment entropy and lattice entropy, following the approach of Derrida et al. [95].

Corollary 3.1 *Let $s = \mathrm{ess}\,\sup_{\mathbb{P}}\omega(t,x)$.*
The function $\lambda_2(\beta)$ is increasing on $\mathbb{R}_+$, with $\lambda_2(+\infty) = -\ln \mathbb{P}(\omega(t,x) = s)$. Thus, the condition (L2) (3.12) holds for all $\beta \geq 0$ as soon as $\mathbb{P}(\omega(t,x) = s) > \pi_d$.

Proof Let q be the law of $\omega(t,x)$. In view of Theorem 3.3, it is enough to show that

$$\lambda_2(\beta) \xrightarrow{\beta \nearrow \infty} \begin{cases} \infty, & \text{if } s = \infty \\ -\ln q(\{s\}) & \text{if } s < \infty. \end{cases} \tag{3.15}$$

The claim is clearly true if $s = \infty$. On the other hand, we observe that, when $q(\{s\}) > 0$,

$$\lambda(\beta) = \beta s + \ln q(\{s\}) + \varepsilon(\beta), \tag{3.16}$$

for some $\varepsilon(\beta) \to 0$ as $\beta \to +\infty$. This gives directly

$$\lambda_2(\beta) = \lambda(2\beta) - 2\lambda(\beta) \to -\ln q(\{s\})$$

as $\beta \to \infty$. To show (3.16), we can write for $h > 0$,

$$\mathbb{P}(e^{\beta\omega}; \omega = s) \le \mathbb{P}(e^{\beta\omega}) = \mathbb{P}(e^{\beta\omega}; \omega \in [s-h, s]) + \mathbb{P}(e^{\beta\omega}; \omega \le s - h),$$

which yields the bounds

$$\beta s + \ln q(\{s\}) \le \lambda(\beta) \le \beta s + \ln q(\{s\}) + \ln\left(\frac{q([s-h, s]) + e^{-\beta h}}{q(\{s\})}\right).$$

Observing that $\inf\{\frac{q([s-h,s])+e^{-\beta h}}{q(\{s\})}; h > 0\}$ decreases to 1 as $\beta \to +\infty$, we obtain (3.16). $\qquad\square$

Example 3.1 Gaussian environment. If ω is standard gaussian $\mathcal{N}(0, 1)$, then $\lambda_2(\beta) = \beta^2$ and hence (3.12) holds if $\beta < \sqrt{\ln(1/\pi_d)}$.

Example 3.2 Absence of strong disorder regime. Consider the case of Bernoulli environment, where $\omega(t, x) = 1$ or 0 with probability p and $1 - p$ respectively. By Corollary 3.1, (3.12) holds for all $\beta \ge 0$ if $p > \pi_d$. Theorem 3.3 shows that, in this case, weak disorder holds for all $\beta \ge 0$.

We call L^2 **region**, the set of parameters β such that (3.12) holds. In this region, the natural martingale is bounded in L^2, and it allows second moment computations. Therefore, a number of results are known, we will see some in the next sections. As observed below in Theorem 3.3, the intersection of the L^2 **region** with $(0, +\infty)$ is an interval $(0, \beta_1)$ with some $\beta_1 \in [0, \infty]$.

3.3   Diffusive Behavior in L^2 Region

All through this section we assume that $d \ge 3$, and that β belongs to the L^2 region.

The next theorem states that, in this region, the random environment does not change the transversal fluctuations of the polymer for large d and small enough β.

Theorem 3.4 ([46, 141, 218])   *Under the assumptions of Theorem 3.3, we have*

$$\lim_{n \nearrow \infty} P_n^{\beta,\omega}[|S_n|^2]/n = 1 \quad \mathbb{P}\text{-}a.s. , \tag{3.17}$$

and for all $f \in C(\mathbb{R}^d)$ with at most polynomial growth at infinity

$$\lim_{n \nearrow \infty} P_n^{\beta,\omega} \left[f \left(S_n / \sqrt{n} \right) \right] = (2\pi)^{-d/2} \int_{\mathbb{R}^d} f \left(x / \sqrt{d} \right) \exp(-|x|^2/2) dx, \quad \mathbb{P}\text{-a.s.}$$
(3.18)

In particular, with Z a d-dimensional gaussian vector $Z \sim \mathcal{N}_d(0, d^{-1}I_d)$, we have

$$P_n^{\beta,\omega} \left(\frac{S_n}{\sqrt{n}} \in \cdot \right) \longrightarrow P(Z \in \cdot) \quad \mathbb{P} - a.s.$$

Remark 3.3 The first rigorous proof of (3.17) was obtained by Imbrie and Spencer [141] in the case of Bernoulli environment. The fact that the polymer is diffusive in some regime was much of a surprise. Soon afterwards, a more transparent proof based on the martingale analysis was given by Bolthausen [46]. The martingale proof was then extended to general environment under condition (3.12) by Song and Zhou [218]. The diffusive behavior (3.17) follows from (3.18) by choosing $f(x) = |x|^2$. In [46], (3.18) is obtained for the Bernoulli environment only. However, with the help of the observation made in [218], it is not difficult to extend the central limit theorem to general environment under the assumption in Theorem 3.3. In [8] Albeverio and Zhou proved, under the assumptions of Theorem 3.3, that under the polymer measure $P_n^{\beta,\omega}$, the path S satisfies the invariance principle for almost every realization of the environment.

Proof Warm-up computation: Before starting the proof of Theorem 3.4, we give an elementary proof of (3.18), in a weaker version with convergence in probability instead of almost sure. Writing the gaussian law $\nu = \mathcal{N}_d(0, d^{-1}I_d)$, we will prove that for all bounded continuous function $g : \mathbb{R} \to \mathbb{R}$,

$$P_n^{\beta,\omega} \left[g \left(n^{-1/2} S_n \right) \right] \to \nu(g)$$
(3.19)

in $\mathbb{P}$-probability as $n \to \infty$. We let $\nu_n(\cdot) = P[n^{-1/2}S_n \in \cdot]$. By the central limit theorem, $\nu_n \to \nu$ weakly as $n \to \infty$.

$$\mathbb{P} \left(\left| P_n^{\beta,\omega} \left[g \left(n^{-1/2} S_n \right) \right] - \nu_n(g) \right|^2 W_n^2 \right)$$

$$= P^{\otimes 2} \mathbb{P} \left(e^{\beta H_n(S) + \beta H_n(\tilde{S}) - 2n\lambda} \left[g \left(n^{-1/2} S_n \right) - \nu_n(g) \right] \left[g \left(n^{-1/2} \tilde{S}_n \right) - \nu_n(g) \right] \right)$$

$$= P^{\otimes 2} \left(e^{\lambda_2 N_n} \left[g \left(n^{-1/2} S_n \right) - \nu_n(g) \right] \left[g \left(n^{-1/2} \tilde{S}_n \right) - \nu_n(g) \right] \right)$$
(3.20)

We know that, under $P^{\otimes 2}$, the r.v. N_n converges to N_∞ a.s., and that $n^{-1/2}S_n$—and similarly $n^{-1/2}\tilde{S}_n$—converges to ν in law. Now, we claim that, under $P^{\otimes 2}$, the triple

$$(N_n, n^{-1/2}S_n, n^{-1/2}\tilde{S}_n) \xrightarrow{\text{law}} (N, Z, \tilde{Z})$$
(3.21)

with $(N, Z, \tilde{Z})$ an independent triple where N has the same law as N_∞, Z and $\tilde{Z}$ have the law v. The proof of this fact makes use of the observation that

$$\sup_{n \geq m} P^{\otimes 2}(N_n \neq N_m) \to 0 \quad \text{as} \quad m \to \infty$$

since $N_n \nearrow N_\infty < \infty$ a.s. Fix $m \geq 1$ and $f, g, \tilde{g}$ continuous and bounded. For all $n \geq m$, we write

$$P^{\otimes 2}\left[f(N_n)g\big(n^{-1/2}S_n\big)\tilde{g}\big(n^{-1/2}\tilde{S}_n\big)\right]$$
$$= P^{\otimes 2}\left[f(N_n)g\big(n^{-1/2}S_n\big)\tilde{g}\big(n^{-1/2}\tilde{S}_n\big)\mathbf{1}_{N_n=N_m}\right] + \varepsilon(n, m)$$
$$= P^{\otimes 2}\left[f(N_m)g\big(n^{-1/2}S_n\big)\tilde{g}\big(n^{-1/2}\tilde{S}_n\big)\mathbf{1}_{N_n=N_m}\right] + \varepsilon(n, m)$$
$$= P^{\otimes 2}\left[f(N_m)g\big(n^{-1/2}(S_n - S_m)\big)\tilde{g}\big(n^{-1/2}(\tilde{S}_n - \tilde{S}_m)\big)\mathbf{1}_{N_n=N_m}\right] + \varepsilon'(n, m)$$
$$= P^{\otimes 2}\left[f(N_m)g\big(n^{-1/2}(S_n - S_m)\big)\tilde{g}\big(n^{-1/2}(\tilde{S}_n - \tilde{S}_m)\big)\right] + \varepsilon''(n, m)$$
$$= P^{\otimes 2}[f(N_m)] \times P\left[g\big(n^{-1/2}(S_n - S_m)\big)\right] \times P\left[\tilde{g}\big(n^{-1/2}(\tilde{S}_n - \tilde{S}_m)\big)\right] + \varepsilon''(n, m) ,$$

which equalities define the terms $\varepsilon(n, m)$, $\varepsilon'(n, m)$, $\varepsilon''(n, m)$ on their first occurrence. Here,

$$|\varepsilon(n, m)| \leq \|f\|_\infty \|g\|_\infty \|\tilde{g}\|_\infty P(N_n \neq N_m)$$

tends to 0 as $m \to \infty$ uniformly in $n \geq m$, $\varepsilon'(n, m) - \varepsilon(n, m) \to 0$ as $n \to \infty$ for all fixed m, and $\sup_{n \geq m} \varepsilon''(n, m) \to 0$ as $m \to \infty$. The last equality comes from independence in the increments of the random walks, and of the two random walks S and $\tilde{S}$. Hence, letting $n \to \infty$ and then $m \to \infty$, we get

$$P^{\otimes 2}\left[f(N_n)g\big(n^{-1/2}S_n\big)\tilde{g}\big(n^{-1/2}\tilde{S}_n\big)\right] \to P^{\otimes 2}[f(N_\infty)] \times v[g] \times v[\tilde{g}]$$

which proves (3.21). Coming back to (3.20), the convergence in law (3.21) will imply that

$$\mathbb{P}\left(\left|P_n^{\beta,\omega}\left[g\left(n^{-1/2}S_n\right)\right] - v_n(g)\right|^2 W_n^2\right) \to P^{\otimes 2}(e^{\lambda_2 N_\infty})\left[v(g) - v(g)\right]^2 = 0$$

since $P^{\otimes 2}(e^{\gamma N_n}) < \infty$ for some (small enough) $\gamma > \lambda_2$. Indeed, by Skorokhod representation theorem, a sequence which converges in law can be constructed on some probability space in order to have almost sure convergence; then, with the uniform integrability condition $P^{\otimes 2}(e^{\gamma N_n}) < \infty$, we obtain convergence in L^1.

Since W_n^{-2} converges to a finite limit, it is bounded in probability, so the previous limit yields (3.19). $\qquad\square$

We now start the proof of Theorem 3.3. The proofs are based on the L^2 analysis of certain martingales on $(\Omega, \mathscr{G}, \mathbb{P})$. This approach was introduced by Bolthausen

[46] and later taken up by Song and Zhou [218]. We summarize the main technical step of their analysis in Proposition 3.2 below.

Some New Martingales We will construct a family of martingales $(M_n)_{n \geq 1}$ on $(\Omega, \mathscr{G}, \mathbb{P})$ of the form:

$$M_n = P[\varphi(n, S_n)\bar{\zeta}_n]. \tag{3.22}$$

Here, $\bar{\zeta}_n$ has been introduced in (3.2) and $\varphi : \mathbb{N} \times \mathbb{Z}^d \to \mathbb{R}$ is a function for which we assume the following properties:

(P1) There are constants $C_i, p \in \mathbb{N}$, $i = 0, 1, 2$ such that

$$|\varphi(n, x)| \leq C_0 + C_1 |x|^p + C_2 n^{p/2} \quad \text{for all } (n, x) \in \mathbb{N} \times \mathbb{Z}^d. \tag{3.23}$$

(P2) $\Phi_n \stackrel{\text{def.}}{=} \varphi(n, S_n)$, $n \geq 1$ is a martingale on $(\Omega_{\text{traj}}, \mathscr{F}, P)$ with respect to the filtration

$$\mathscr{F}_n = \sigma[S_j ; j \leq n]. \tag{3.24}$$

It is easy to see that $(M_n)_{n \geq 1}$ is a $(\mathscr{G}_n)$-martingale on $(\Omega_{\text{traj}}, \mathscr{G}, \mathbb{P})$: Denoting conditional expectations by $\mathbb{P}^{\mathscr{G}_n}, P^{\mathscr{F}_n}$, we have

$$
\begin{aligned}
\mathbb{P}^{\mathscr{G}_n} M_{n+1} &= P[\varphi(n+1, S_{n+1})\bar{\zeta}_n \mathbb{P}^{\mathscr{G}_n} e^{\beta\omega(n+1, S_{n+1})-\lambda}] \\
&= P[\varphi(n+1, S_{n+1})\bar{\zeta}_n] \\
&= P[\bar{\zeta}_n P^{\mathscr{F}_n} \varphi(n+1, S_{n+1})] \\
&= M_n,
\end{aligned} \tag{3.25}
$$

by (P2). The following proposition generalizes [46, Lemma 4] and [218, Theorem 2].

Proposition 3.2 *Consider the martingale $(M_n)_{n \geq 1}$ defined by (3.22). Suppose that $d \geq 3$ and that (3.12), (P1), (P2) are satisfied. Then, there exists $\kappa \in [0, p/2)$ such that*[1]

$$\max_{0 \leq j \leq n} |M_j| = \mathcal{O}(n^\kappa), \quad \text{as } n \nearrow \infty, \mathbb{P}\text{-a.s.} \tag{3.26}$$

If in addition, $p < \frac{1}{2}d - 1$, then

$$\lim_{n \nearrow \infty} M_n \text{ exists } \mathbb{P}\text{-a.s. and in } L^2(\mathbb{P}). \tag{3.27}$$

---

[1]This means that $\sup_n n^{-\kappa} \max_{0 \leq j \leq n} |M_j| < \infty$ a.s.

Remark 3.4

(i) Note that Theorem 3.3 can be viewed as a consequence of (3.27) by choosing $\varphi \equiv 1$, in which case $M_n = W_n$ and $p = 0$.

(ii) As will be seen from the way (3.26) is used below, it is crucial that the divergence of the right-hand-side is strictly slower than $n^{p/2}$, and this is where the property (P2) is relevant. If we drop the property (P2) from the assumption of Proposition 3.2, we then have a larger bound:

$$M_n = \mathcal{O}(n^{p/2}), \quad \text{as } n \nearrow \infty, \mathbb{P}\text{-a.s.} \tag{3.28}$$

This larger bound from the weaker assumption can be obtained via Proposition 3.2 as in the proof of (2.1) in [46].

Proof (Proof of Theorem 3.3) We will prove Proposition 3.2 later on, independently. With Proposition 3.2 in hand, we first derive the statements in Theorem 3.3, following the lines of Bolthausen [46].

- To prove (3.17), we take $\varphi(n,x) = |x|^2 - n$ (hence $p = 2$). Then, by Theorem 3.3 and Proposition 3.2, there exists $\kappa \in [0,1)$ such that

$$P_n^{\beta,\omega}[|S_n|^2] - n = P[\varphi(n, S_n)\bar{\zeta}_n]/W_n = \mathcal{O}(n^\kappa) \quad \mathbb{P}\text{-a.s.} \tag{3.29}$$

- We now explain the route to (3.18).

We let $a = (a_j)_{j=1}^d$ and $b = (b_j)_{j=1}^d$ denote multi indices in what follows. We will use standard notation $|a|_1 = a_1 + \ldots + a_d$, $x^a = x_1^{a_1} \cdots x_d^{a_d}$ and $(\frac{\partial}{\partial x})^a = \left(\frac{\partial}{\partial x_1}\right)^{a_1} \cdots \left(\frac{\partial}{\partial x_d}\right)^{a_d}$ for $x \in \mathbb{R}^d$. It is enough to prove (3.18) for any monomial of the form $f(x) = x^a$. We will do this by induction on $|a|_1$. We introduce

$$\varphi(n,x) = \left(\frac{\partial}{\partial\theta}\right)^a \exp(\theta \cdot x - n\rho(\theta))\Big|_{\theta=0},$$

$$\psi(n,x) = \left(\frac{\partial}{\partial\theta}\right)^a \exp\left(\theta \cdot x - n\frac{|\theta|^2}{2d}\right)\Big|_{\theta=0},$$

where $\rho(\theta) = \ln\left(\frac{1}{d}\sum_{1\le j\le d}\cosh(\theta_j)\right)$. Clearly, the function φ satisfies (P1) and (P2) with $p = |a|_1$. On the other hand, we see from the definition of ψ that

$$(2\pi)^{-d/2} \int_{\mathbb{R}^d} \psi(1, x/\sqrt{d})e^{-|x|^2/2}dx = 0. \tag{3.30}$$

Moreover, it is not difficult to see [46, Lemma 3c] that $\varphi(n,x) = x^a + \varphi_0(n,x)$ and $\psi(n,x) = x^a + \psi_0(n,x)$ where

$$\varphi_0(n,x) = \sum_{\substack{|b|_1+2j\leq|a|_1 \\ j\geq 1}} A_a(b,j)x^b n^j, \quad \psi_0(n,x) = \sum_{\substack{|b|_1+2j=|a|_1 \\ j\geq 1}} A_a(b,j)x^b n^j.$$

for some $A_a(b,j) \in \mathbb{R}$. In particular, φ_0 and ψ_0 have the same coefficients for $x^b n^j$ with $|b|_1 + 2j = |a|_1$. We now write

$$\left(x/\sqrt{n}\right)^a = \varphi(n,x)n^{-|a|_1/2} - \psi_0(1,x/\sqrt{n}) + \left[\psi_0(n,x) - \varphi_0(n,x)\right]n^{-|a|_1/2},$$

which yields a decomposition for $P_n^{\beta,\omega}[(S_n/\sqrt{n})^a]$ as

$$P_n^{\beta,\omega}[(S_n/\sqrt{n})^a] = \frac{1}{W_n}P[\varphi(n,S_n)\bar{\zeta}_n]n^{-|a|_1/2} - \frac{1}{W_n}P[\psi_0(1,S_n/\sqrt{n})\bar{\zeta}_n]$$

$$+ \frac{1}{W_n}P[(\psi_0(n,S_n) - \varphi_0(n,S_n))\,\bar{\zeta}_n]n^{-|a|_1/2}$$

As $n \nearrow \infty$, the second term in the right-hand side converges to $(2\pi)^{-d/2}\int_{\mathbb{R}^d}(x/\sqrt{d})^a e^{-|x|^2/2}dx$ by the induction hypothesis and (3.30). The first and the third terms vanish as $n \nearrow \infty$. In fact, we use Theorem 3.3, Proposition 3.2 for the first term and Theorem 3.3, (3.28) for the third term. $\quad\square$

We now turn to the proof of Proposition 3.2, following [218].

Proof Main steps:

- A tedious computation (see [79]) shows that

$$\mathbb{P}[M_n^2] = \mathscr{O}(b_n), \qquad b_n = \sum_{1\leq j\leq n} j^{p-\frac{d}{2}}. \tag{3.31}$$

- Setting $M_n^* = \max_{0\leq j\leq n}|M_j|$. For (3.26) it is sufficient to prove that for any $\delta > 0$,

$$M_n^* = \mathscr{O}(n^\delta\sqrt{b_n}) \quad \text{as } n \nearrow \infty, \mathbb{P}\text{-a.s.} \tag{3.32}$$

Moreover, by the monotonicity of M_n^* and the polynomial growth of $n^\delta\sqrt{b_n}$, it is enough to prove (3.32) along a subsequence $\{n^k : n \geq 1\}$ for some power $k \geq 2$. Now, take $k > 1/\delta$. We then have by Chebychev's inequality, Doob's inequality and (3.31) that

$$\mathbb{P}\{M^*_{n^k} > n^{k\delta}\sqrt{b_{n^k}}\} \leq \mathbb{P}\{M^*_{n^k} > n\sqrt{b_{n^k}}\}$$
$$\leq \mathbb{P}[(M^*_{n^k})^2]/(n^2 b_{n^k})$$
$$\leq 4\mathbb{P}[M^2_{n^k}]/(n^2 b_{n^k})$$
$$\leq Cn^{-2}.$$

Then, it follows from the Borel-Cantelli lemma that

$$\mathbb{P}\{M^*_{n^k} \leq n^{k\delta}\sqrt{b_{n^k}} \text{ for large enough } n's\} = 1.$$

This ends the proof of (3.26). The second statement (3.27) in Proposition 3.2 follows from (3.31) and the martingale convergence theorem. This completes the proof of Proposition 3.2. □

Remark 3.5 The parameters—mean and covariance—in the Central Limit Theorem 3.3 do not depend on the environment. On the contrary, the density in the LLT does depend on the environment as seen from the final point. Further, the first order corrections do depend on the particular realization of the environment. Indeed, for $d \geq 7$ the statement in (3.29) can be refined into

$$P^{\beta,\omega}_n(|S_n|^2) - n \quad \text{converges a.s. and in } L^2$$

in view of (3.27) and of $p < d/2 - 1$ with $p = 2, d \geq 7$. It is easy to see that the limit is not a constant, and therefore

$$P^{\beta,\omega}_n(|S_n|^2) = n + V + o(1)$$

as $n \to \infty$, where $V = V(\omega)$ is a non trivial function of the environment. Similarly for the first moment, we can use (3.27) for $d \geq 5$ since $p = 1$ in this case, and we conclude that $P^{\beta,\omega}_n(S_n)$ converges to a random vector as $n \to \infty$.

The first order corrections are studied in details in [45], where they are given in a rather explicit form in terms of a perturbation expansion.

3.4   Local Limit Theorem in the L^2 Region

In this section we will consider the point-to-point (normalized) partition function

$$W_n(y) = P\left[\exp\{\beta H_n(S) - n\lambda\}\Big| S_n = y\right], \quad y \in \mathbb{Z}^d. \tag{3.33}$$

The terminology is easily understood, indicating that the endpoint is fixed at (n, y). In contrast, the endpoint at time n in (3.1) is free, and the corresponding partition function is called point-to-level (or point-to-line in dimension $d = 1$). Traditional abbreviations are P2P and P2L respectively.

When $n \in \mathbb{N}^*$ and $x \in \mathbb{Z}^d$ are such that $P(S_n = x) > 0$—that is, when $|x|_1 \le n$ and $|x|_1 = n$ modulo 2—, we write

$$n \leftrightarrow x \iff P(S_n = x) > 0 \tag{3.34}$$

In this case, the random variable $W_n(x)$ has mean 1, and we have

$$W_n = \sum_y W_n(y) P(S_n = y).$$

Define $\theta_{n,x}^{\leftarrow}$ the reflection operator on the environment, given by

$$\theta_{n,x}^{\leftarrow}(\omega) = \omega' \quad \text{with} \quad \omega'(u, y) = \omega(n - u, x + y).$$

In this section, for $0 \le k \le n$ we will use the notations

$$\bar{\zeta}_{k,n}(\mathbf{x}) = \exp\left\{\beta \sum_{i=k}^{n} \omega(i, x_i) - (n - k + 1)\lambda(\beta)\right\},$$

so that $\bar{\zeta}_{1,n} = \bar{\zeta}_n$, and

$$\bar{\zeta}_{k,n}^{\leftarrow}(\mathbf{x}) = \exp\left\{\beta \sum_{i=k-1}^{n-1} \omega(n - i, x_i) - (n - k + 1)\lambda(\beta)\right\},$$

where we note the changes in the indices. In particular,

$$P^y(\bar{\zeta}_{1,n}^{\leftarrow}) \overset{\text{law}}{=} W_n.$$

Observe that, the random variable $W_\infty \circ \theta_{n,x}^{\leftarrow}$ is well defined for all n, x, and that

$$P^y(\bar{\zeta}_{1,n}^{\leftarrow}) - W_\infty \circ \theta_{n,x}^{\leftarrow} \longrightarrow 0$$

in probability (and in L^2-norm in the L^2 region) by the above equality in law. We now state the main result.

Theorem 3.5 (Local Limit Theorem) *Assume (3.12). Then, for all $A < \infty$ and all sequence of integers $\ell_n \to \infty$ with $\ell_n = o(n^\alpha)$ for some $\alpha < 1/2$, we have*

$$P\left[\bar{\zeta}_{1,n} \big| S_n = y\right] = P(\bar{\zeta}_{1,\ell_n}) P^y(\bar{\zeta}_{1,\ell_n}^{\leftarrow}) + \delta_n^y, \tag{3.35}$$

with

$$\lim_{n\to\infty} \sup\{\mathbb{P}(|\delta_n^y|^2); |y| \le An^{1/2}, n \leftrightarrow y\} = 0.$$

Moreover,

$$W_n(y) = W_\infty \times (W_\infty \circ \theta_{n,y}^{\leftarrow}) + \varepsilon_n^y, \tag{3.36}$$

where the error term $\varepsilon_n^y \to 0$ *in* L^1 *uniformly on* $\{y : |y| \le An^{1/2}, n \leftrightarrow y\}$.

Remark 3.6

(i) The result goes back to Sinai [216], and was extended to more general models in [234].

(ii) This result means that the polymer measure at time n large, is close to the gaussian measure up to a factor which depends on the environment around the final point. We don't give a formal result, but instead we can write loosely

$$P_n^{\beta,\omega}(S_n = y) \simeq W_\infty \circ \theta_{n,y}^{\leftarrow} \times P(S_n = y). \tag{3.37}$$

Recall at this point that W_∞ has expectation 1 under the above condition.

(iii) Intuitively, the polymer only feels the environment in the neighborhood of the starting and ending points, and behave like gaussian in between. The result is the counterpart in positive temperature of the following result for supercritical Bernoulli percolation in Durrett's book [103, Exercise P.201]: The minimal number of closed bonds one has to cross when traveling from the origin to a distant point (n, y) converges to the sum of two independent copies of the distance from the origin to the infinite cluster.

Proof Main steps from of the proof of Theorem 3.5 [234]:

Step 1: with $\ell_n \le n/2$, we first approximate $W_n(y)$ with $P\left[\bar{\zeta}_{1,\ell_n} \bar{\zeta}_{n-\ell_n,n} | S_n = y\right]$ in L^2.

$$\lim_{n\to\infty} \sup_{|y|\le An^{1/2}} \left\| W_n(y) - P\left[\bar{\zeta}_{1,\ell_n} \bar{\zeta}_{n-\ell_n,n} | S_n = y\right] \right\|_2^2 = 0.$$

This holds because, by transience, two random walk paths starting and ending at the same points will have no intersections between times ℓ_n and $n - \ell_n$ w.h.p.

Step 2: Estimating the second moment, it can be seen that

$$\lim_{n\to\infty} \left\| P\left[\bar{\zeta}_{1,\ell_n} \bar{\zeta}_{n-\ell_n,n} | S_n = y\right] - P[\bar{\zeta}_{1,\ell_n}] P^y[\bar{\zeta}_{1,\ell_n}^{\leftarrow}] \right\|_1 = 0$$

using the standard local limit theorem for the simple random walk. $\qquad\square$

3.5 Analytic Functions Method: Energy and Entropy in the L^2 Region

In the L^2 region, we can also consider the fluctuations of extensive thermodynamic quantities other than the partition function: we show that these are typically of order 1, like $\ln Z_n(\omega; \beta)$ itself. Assume (3.12), and consider the two following variables: the energy averaged over the path and the relative entropy of $P_n^{\beta,\omega}$ with respect to P, as defined in (3.41). We will show that

$$P_n^{\beta,\omega}[H_n] = n\lambda'(\beta) + \mathcal{O}(1) , \quad \mathcal{H}(P_n^{\beta,\omega}|P) = n[\beta\lambda'(\beta) - \lambda(\beta)] + \mathcal{O}(1) ,$$

where the remainders $\mathcal{O}(1)$ converge to a limit that we identify. The proof uses analytic functions arguments. We will view $W_n = W_n(\beta)$ as a function of β, and, in this section, we will write explicitly the dependence in the parameter β. The martingales $W_n(\beta)$ are analytic in β, and the a.s. convergence as $n \to \infty$ is shown, in the L^2-region, to be a convergence in the sense of analytic functions. The crucial estimate is a bound on the second moment of some complex random variable, this explains why we do assume (3.12).

An Analytic Family of Martingales For β complex, $\mathbb{P}[\exp \beta\omega(n,x)]$ is well defined, but we also want its logarithm to be holomorphic. Let U_0 be the open set in the complex plane given by

$$U_0 = \text{connected component of 0 in } \{\beta \in \mathbb{C}; \ \mathbb{P}[\exp \beta\omega(n,x)] \notin \mathbb{R}_-\} .$$

Then, U_0 is a neighborhood of the real axis by continuity. By composition with the complex logarithm log (in its principal determination), the function $\lambda(\beta) = \log \mathbb{P}[\exp\{\beta\omega(t,x)\}]$ is analytic on U_0. Define, for $n \geq 0$ and $\beta \in U_0$,

$$W_n(\beta) = P\left[\exp \left(\beta \sum_{t=1}^n \omega(t, S_t) - n\lambda(\beta) \right) \right]. \tag{3.38}$$

Then, for all $\beta \in U_0$, the sequence $(W_n(\beta), n \geq 0)$ is a $(\mathscr{G}_n)_n$-martingale with complex values, where $\mathscr{G}_n = \sigma\{\omega(t,x); t \leq n, x \in \mathbb{Z}^d\}$. At the same time, for each n and ω, $W_n(\beta)$ is an analytic function of $\beta \in U_0$.

In view of (3.12), consider the subset U_1 of real numbers

$$U_1 = \{\beta \in \mathbb{R} \ : \ \lambda(2\beta) - 2\lambda(\beta) < -\ln \pi_d\} , \tag{3.39}$$

defined by condition L^2. We know that $W_\infty > 0$ a.s. for $\beta \in U_1$, and that U_1 is an open interval (β_1^-, β_1^+) containing 0 $(-\infty \leq \beta_1^- < 0 < \beta_1^+ \leq +\infty)$. The following proposition is the main technical tool, it was established in [75].

Proposition 3.3 *Define U_2 to be the connected component of 0 in the set*

$$\left\{ \beta \in U_0 \;:\; \lambda(2\Re\beta) - 2\Re\lambda(\beta) < -\ln\pi_d \right\}$$

which contains the origin. Then, U_2 is a complex neighborhood of U_1, and $\sup_n \mathbb{P}(|W_n|^2)$ is finite for all $\beta \in U_2$. Furthermore, there exists an event Ω_1 with $\mathbb{P}(\Omega_1) = 1$ such that the following limit exists:

$$W_\infty(\beta) = \lim_{n\to\infty} W_n(\beta), \quad \text{for all } \omega \in \Omega_1, \beta \in U_2 .$$

The convergence is locally uniform. In particular, the limit $W_\infty(\beta)$ is holomorphic in U_2, and all derivatives of W_n converge locally uniformly to the corresponding ones of W_∞, i.e.,

$$\frac{d^k}{d\beta^k} W_n(\beta) \to \frac{d^k}{d\beta^k} W_\infty(\beta) , \qquad \mathbb{P}-a.s.,$$

uniformly on compacts of U_2 ($k \geq 0$). Finally, $W_\infty(\beta) > 0$ for all $\beta \in U_1$, $\mathbb{P}$-a.s.

Proof The set U_2 is open by continuity, and it contains U_1. Since $\overline{(e^z)} = e^{\bar{z}}$ and $\overline{\mathbb{P}[f]} = \mathbb{P}[\bar{f}]$, we have $\overline{\lambda(\beta)} = \lambda(\bar{\beta})$, and further

$$\mathbb{P}\left(e^{\beta\omega(t,x) + \bar{\beta}\omega(t,y)} \right) = \begin{cases} e^{\lambda(\beta) + \lambda(\bar{\beta})} = e^{2\mathrm{Re}\lambda(\beta)}, & \text{if } x \neq y, \\ e^{\lambda(\beta + \bar{\beta})} = e^{\lambda(2\mathrm{Re}\beta)}, & \text{if } x = y. \end{cases}$$

Hence,

$$\mathbb{P}\left[|W_n(\beta)|^2 \right] = \mathbb{P}\left[P[\exp\{\beta H_n(S) - n\lambda(\beta)\}] P[\exp\{\bar{\beta}H_n(\tilde{S}) - n\overline{\lambda(\beta)}\}] \right]$$

$$= P^{\otimes 2}\left[\mathbb{P}\left[\exp\{\beta H_n(S) + \bar{\beta}H_n(\tilde{S}) - 2n\mathrm{Re}\lambda(\beta)\} \right] \right]$$

$$= P^{\otimes 2}\left[\exp\left\{ [\lambda(2\mathrm{Re}\beta) - 2\mathrm{Re}\lambda(\beta)] \sum_{t=1}^{n} \mathbf{1}\{S_t = \tilde{S}_t\} \right\} \right]$$

$$\leq P^{\otimes 2}\left[\exp\left\{ [\lambda(2\mathrm{Re}\beta) - 2\mathrm{Re}\lambda(\beta)] \sum_{t=1}^{\infty} \mathbf{1}\{S_t = \tilde{S}_t\} \right\} \right]$$

$$< \infty \tag{3.40}$$

if $\beta \in U_2$. Indeed, the random variable $\sum_{t=1}^{\infty} \mathbf{1}\{S_t = \tilde{S}_t\}$ (which is the number of meetings between two independent d-dimensional simple random walks) is geometrically distributed with parameter π_d.

For any *real* $\beta \in U_2$, the positive martingale $W_n(\beta)$ is bounded in L^2, hence it converges almost surely and in L^2-norm to a positive limit $W_\infty(\beta)$.

We need a stronger convergence result. Fix a point $\beta \in U_2$ and a radius $r > 0$ such that the closed disk $D(\beta, r) \subset U_2$. Choosing $R > r$ such that $D(\beta, R) \subset U_2$, we obtain by Cauchy's integral formula for all $\beta' \in D(\beta, r)$,

$$W_n(\beta') = \frac{1}{2i\pi} \int_{\partial D(\beta,R)} \frac{W_n(z)}{z - \beta'} dz = \int_0^1 \frac{W_n(\beta + Re^{2i\pi u})Re^{2i\pi u}}{(\beta + Re^{2i\pi u}) - \beta'} du ,$$

hence

$$X_n := \sup\{|W_n(\beta')|; \beta' \in D(\beta, r)\} \leq R \int_0^1 \frac{|W_n(\beta + Re^{2i\pi u})|}{R - r} du .$$

Letting $C = (R/(R - r))^2$, we obtain by the Schwarz inequality

$$(\mathbb{P}[X_n])^2 \leq C\mathbb{P}[\int_0^1 |W_n(\beta + Re^{2i\pi u})|^2 du]$$

$$\leq C \sup\{\mathbb{P}[|W_n(\beta'')|^2]; n \geq 1, \beta'' \in D(\beta, R)\}$$

$$< \infty$$

in view of (3.40). Notice now that X_n, a supremum of positive submartingales, is itself a positive submartingale. Since $\sup_n \mathbb{P}[X_n] < \infty$, X_n converges $\mathbb{P}$-a.s. to a finite limit X_∞. Finally,

$$\sup\{|W_n(\beta')|; \beta' \in D(\beta, r), n \geq 1\} = \sup_n X_n < \infty \quad \mathbb{P}\text{-a.s.,}$$

and W_n is uniformly bounded on compact subsets of U_2 on a set of environments of full probability. On this set, $(W_n, n \geq 0)$ is a normal sequence [204] which has a unique limit on the real axis: since U_2 is connected, the full sequence converges to some limit W_∞, which is holomorphic on U_2, and, as mentioned above, positive on the real axis. By standard results of convergence of analytic functions, the derivatives also converge. $\qquad\square$

Note that we do not know that $W_\infty(\beta) \neq 0$ for general $\beta \in U_2$, except for $\beta \in U_1$—and of course for some complex neighborhood around U_1—. We draw now some consequences for real β's.

Recall the (relative) entropy $\mathcal{H}(\nu|\mu)$ of two probability measures , ν, μ defined on the same space:

$$\mathcal{H}(\nu|\mu) = \begin{cases} \nu(\ln \frac{d\nu}{d\mu}) \text{ if } \nu \ll \mu \text{ and } \ln \frac{d\nu}{d\mu} \in L^1(\nu) \\ +\infty \quad \text{otherwise} \end{cases} \tag{3.41}$$

Theorem 3.6 ([75]) *Let $d \geq 3$. Then, we have:*

(i) W_∞ and $\ln W_\infty$ are analytic (real) function of $\beta \in U_1$.
(ii) For all $\beta \in U_1$, as $n \to \infty$, the internal energy is such that

$$P_n^{\beta,\omega}[H_n] - n\lambda'(\beta) \to (\ln W_\infty)'(\beta) , \tag{3.42}$$

and the relative entropy,

$$\mathscr{H}(P_n^{\beta,\omega}|P) - n[\beta\lambda'(\beta) - \lambda(\beta)] \to \beta(\ln W_\infty)'(\beta) - \ln W_\infty(\beta) . \tag{3.43}$$

(iii) On the other hand, for all $\beta \in U_1$ and $\mathbb{P}$-a.e. environment, the law of $\dfrac{H_n - n\lambda'(\beta)}{\sqrt{n}}$ under $P_n^{\beta,\omega}$ converges to the Gaussian $\mathscr{N}(0, \lambda''(\beta))$ where $\lambda''(\beta) > 0$.

Proof We have the identities

$$(\ln W_n)'(\beta) = P_n^{\beta_2,\omega}[H_n] - n\lambda'(\beta),$$
$$\mathscr{H}(P_n^{\beta_2,\omega}|P) = \beta P_n^{\beta_2,\omega}[H_n] - n\lambda'(\beta) - \ln W_n(\beta) .$$

In view of Proposition 3.3, for $\beta \in U_1$, we have almost sure convergence

$$(\ln W_n)'(\beta) = (W_n)'(\beta)/W_n(\beta) \to (W_\infty)'(\beta)/W_\infty(\beta) = (\ln W_\infty)'(\beta),$$

which is the first result (3.42). The second one (3.43) follows easily. In order to prove the last one, we show the stronger statement that, for $\mathbb{P}$-a.e. environment,

$$P_n^{\beta,\omega}\left[\exp\{\frac{u(H_n - n\lambda'(\beta))}{\sqrt{n}}\}\right] \to \exp\{\frac{\lambda''(\beta)u^2}{2}\}$$

as $n \to \infty$ for all $u \in \mathbb{R}$ and $\beta \in U_2$. Write the left-hand side as

$$\frac{W_n(\beta + un^{-1/2})}{W_n(\beta)} \times \exp\left\{n[\lambda(\beta + un^{-1/2}) - \lambda(\beta) - un^{-1/2}\lambda'(\beta)]\right\}$$

Since $W_n \to W_\infty$ locally-uniformly on U_1, and since λ is smooth, the right-hand side converges $\mathbb{P}$-a.s. to

$$[W_\infty(\beta)/W_\infty(\beta)] \times \exp\{\lambda''(\beta)u^2/2\} = \exp\{\lambda''(\beta)u^2/2\}$$

as $n \to \infty$. $\qquad\qquad\qquad\qquad\qquad\qquad\qquad\qquad\qquad\qquad\qquad\qquad\qquad\square$

Remark 3.7 The *average energy* for the polymer measure, $P_n^{\beta,\omega}[H_n]$, behaves at first order like $n\lambda'(\beta)$ the energy of the annealed model, but it has fluctuations of order one in this part of the weak disorder region. The entropy also has $\mathscr{O}(1)$ fluctuations.

On the other hand, the last result shows that, due to variations from a path to another, the fluctuations of the *energy under the polymer measure* are normal and of order of magnitude $\mathcal{O}(\sqrt{n})$.

Open Problem 3.7 (At Weak Disorder but Outside L2-Region).

(i) Is W_∞ analytic up to β_c?
(ii) Do the LHS of (3.42) and (3.43) converge up to β_c?

3.6 Concentration Revisited

In Theorem 2.2 we give a sub-gaussian fluctuation bound for general models, which implies that $\chi^{\|} \leq 1/2$. All standard methods for concentration inequalities give similar bounds. From simulation it appears that this is not sharp, for any values of d. This phenomenon also occurs for the fluctuations of the top eigenvalue of large random matrices [127]. General methods give symmetric bounds. The lower tail of the large deviations is typically lighter than the upper tail. See [26] for sharp estimates on the lower tail of the large deviations. A complete survey for first passage percolation is [20].

Open Problem 3.8 *Find a general method showing that $\chi^{\|} < 1/2$.*

3.6.1 Inside the L^2 Region

Using Talagrand's concentration techniques [225] and their adaptation to disordered systems [226], a much sharper lower tail inequality has been obtained in the L^2 region. For Gaussian environment see [58, Theorem 1.5]. We state the bounded environment case of [175, Proposition 1]:

Proposition 3.4 *Assume $|\omega(t, x)| \leq 1$ and (3.12). Then, there is some $C > 0$ such that, for all $u > 0, n \geq 1$,*

$$\mathbb{P}[\ln W_n \leq -u] \leq C \exp\left\{-\frac{u^2}{16C\beta^2}\right\}$$

Technically, the bound on the second moment allows to restrict to an event where derivative is small.

3.6.2   General Case: Sublinear Variance Estimate

Till the end of the section, β **is arbitrary**, we cover the general situation, including strong disorder, trying to tackle the above question. Much efforts have been spent, resulting in the small improvement of reducing the variance by a logarithmic factor.

Following the lines of Benjamini, Kalai and Schramm [28] and Benaïm and Rossignol [27] for percolation, the methods described in Theorem 2.2 have been refined [11], leading to concentration of $\ln Z_n(\omega, \beta)$ on scale $(n/\ln n)^{1/2}$. We also mention the estimate from [123] for last passage percolation, based on a coupling argument, which has not been adapted to polymer setup.

Assume the environment has a nearly gamma distribution.[2] Then, for some $c > 0$,

$$\mathbb{P}\left(|\ln Z_n(\omega, \beta) - \mathbb{P}[\ln Z_n(\omega, \beta)]| > t(n/\ln n)^{1/2}\right) \le c^{-1}e^{-ct}, \quad t > 0,$$

and then (Theorem 1.2 in [11]),

$$np(\beta) - C\sqrt{n/\ln n} \times \ln\ln n \le \mathbb{P}[\ln Z_n(\omega, \beta)] \le np(\beta).$$

The first claim is obtained by separating the fluctuations coming from the first steps of the path (in fact, a logarithmic number) and those coming from the rest. The former only uses environment at locations close to the origin. The second estimate is derived from the first one by a coarse graining motivated by Alexander [10], and it is useful when we want to study concentration with respect to the limit instead of the mean.

Viewing $\ln Z_n$ as a (non-linear) mean of H_n over the path space, it is natural to compare the variance of $\ln Z_n$ to the variance of $H_n(\mathbf{x})$ - which does not depend on $\mathbf{x}$. The phenomenon

$$\text{Var}(\ln Z_n) \ll \beta^2 \text{Var}(H_n(\mathbf{x}))$$

(as $n \to \infty$) is known as superconcentration. It appears in many disordered systems, and it has been recently studied from a probabilistic perspective [64]. It relates to the presence of multiple peaks, which are sub-optimal non-overlapping paths. In [99], the authors use the above $n/\ln n$ bound for the variance of $\ln Z_n$ together with general results for superconcentration of Gaussian fields. Fix $\xi, \delta, \epsilon \in (0, 1)$. They obtain, for the polymer in a Gaussian medium, that with probability at least $1 - \xi$, there exists a set of paths of cardinality at least n^C (for some $C = C(\xi, \delta, \epsilon) > 0$) with the two properties: (1) each path has energy H_n within a factor $1 - \delta$ from the expected maximum energy; (2) pairwise correlation are less than ϵ (i.e., $N_n(\mathbf{x}, \mathbf{y}) \le$

---

[2]The main point is that, when $\omega(t, x)$ has a nearly gamma distribution, it satisfies a logarithmic Sobolev inequality w.r.t. a particular differential operator. It covers a wide class of densities, including gamma and Gaussian.

$n\epsilon$ for $\mathbf{x}, \mathbf{y}$ different in A_n). Improving the estimate on the ratio to a polynomial decay (which is not known at the moment, but expected) would increase the number n^c to a stretched exponential.

3.7 Rate of Martingale Convergence

It is natural to determine the rate of convergence of the natural martingale to the limit.

Theorem 3.9 ([70]) *For $d \geq 3$, there exists some $\beta_0 > 0$ such that, for $|\beta| < \beta_0$,*

$$n^{\frac{d-2}{4}} (W_\infty - W_n) \to \sigma_1 W_\infty G \quad \text{in distribution} \tag{3.44}$$

and

$$n^{\frac{d-2}{4}} \frac{(W_\infty - W_n)}{W_n} \to \sigma_1 G \quad \text{in distribution,} \tag{3.45}$$

where

$$\sigma_1^2 = \frac{d^{d/2}(1 - \pi_d)}{2^{d-2}(d-2)\pi^{d/2}\pi_d} \times \text{Var}(W_\infty), \tag{3.46}$$

G is a Gaussian r.v. with law $N(0, 1)$, which is independent of W_∞. Moreover, the convergence in (3.44) is stable, and the convergence in (3.45) is mixing (see the definitions below).

We recall two convergence modes. Let (Y_n) be a sequence of real random variables defined on a common probability space $(\Omega, \mathscr{F}, P)$, converging in distribution to a limit Y.

- This convergence is called *stable* if for all $B \in \mathscr{F}$ with $P(B) > 0$, the conditional law of Y_n given B converges to some probability distribution depending on B.
- This convergence is called *mixing* if it is stable and the limit of conditional laws does not depend on B—and therefore is the law of Y—.

The stable convergence allows to add extra variables: for any fixed r.v. Z on $(\Omega, \mathscr{F}, P)$, the couple (Y_n, Z) converges in law to some coupling of Y and Z on an extended space. For that, convergence in law is not enough; stable convergence is stronger, it requires that the sequence is defined on a common probability space, and tells information on the joint law of the sequence and also with the underlying space. The mixing convergence means that Y_n is asymptotically independent of all event $A \in \mathscr{F}$. These convergences were introduced by Rényi [200]; we refer to [9] for a nice presentation with the main consequences, and to [131, pp. 56–57] for an extended account on the connections with martingale central limit theorem.

Proof (Sketch of Proof) : We apply a version of the central limit theorem for infinite martingale arrays, which is a slight extension of Corollaries 3.1 and 3.2 of the book by Hall and Heyde [131, pp. 58–59 and p. 64], see [70] for more details.

We decompose

$$n^{\frac{d-2}{4}} (W_\infty - W_n) = n^{\frac{d-2}{4}} \sum_{k=n}^{\infty} D_{k+1}, \tag{3.47}$$

where

$$D_{k+1} = W_{k+1} - W_k, \quad k \geq n,$$

forms a sequence of martingale differences with respect to the sequence of filtrations $\mathscr{F}^{(n)} = \left(\mathscr{F}_i^{(n)}\right)_{i \geq 0}$, where $\mathscr{F}_i^{(n)} = \mathscr{G}_{n+i}$. We prove that :

(a) the sequence of series of the conditional variance converges to a random variable:

$$s_n^2 := n^{\frac{d-2}{2}} \sum_{k \geq n} \mathbb{P}[D_{k+1}^2 | \mathscr{G}_k] \rightarrow \sigma_1^2 W_\infty^2 \quad \text{in probability,} \tag{3.48}$$

where $\mathbb{P}(\cdot|\mathscr{G}_k)$ denotes the conditional expectation (or probability) given $\mathscr{G}_k$;

(b) the following Lindeberg condition holds :

$$\forall \varepsilon > 0, \quad n^{\frac{d-2}{2}} \sum_{k \geq n} \mathbb{P}\left[D_{k+1}^2 \mathbf{1}_{\{n^{\frac{d-2}{4}}|D_{k+1}|>\varepsilon\}} | \mathscr{G}_k \right] \rightarrow 0 \quad \text{in probability.} \tag{3.49}$$

From (3.48) and (3.49) we conclude by a version of the martingale central limit theorem that (3.44) and (3.45) hold with the norming $1/W_n$ in (3.45) replaced by $1/W_\infty$. As $W_n/W_\infty \rightarrow 1$ a.s. (and thus in probability), we can change the factor $1/W_\infty$ to $1/W_n$ without changing the convergence in distribution.

To show the convergence (3.48) of the conditional variance, we prove in [70] that there exists $\beta_0 > 0$ such that for $|\beta| < \beta_0$ and σ_1 from (3.46), we have, as $n \rightarrow \infty$,

$$\mathbb{P}[(W_n - W_\infty)^4] \longrightarrow 0, \tag{3.50}$$

$$\mathbb{P}[s_n^4] - \sigma_1^4 \mathbb{P}[W_n^4] \longrightarrow 0, \tag{3.51}$$

$$\mathbb{P}[s_n^2 W_n^2] - \sigma_1^2 \mathbb{P}[W_n^4] \longrightarrow 0. \tag{3.52}$$

For the proof we need to analyse fourth moment of the partition function, looking at four replicas of the walk, and estimating the Green function: To give a flavor, the difference can be written as

$$W_\infty - W_n = \sum_x W_n(x)\big(W_\infty \circ \theta_{n,x} - 1\big), \tag{3.53}$$

and a central quantity is the covariance

$$Cov(W_\infty, W_\infty \circ \theta_{0,x}) = Var(W_\infty)\, P_{0,x}^{\otimes 2}(\exists n \geq 1 : S_n = \tilde{S}_n)$$

$$= Var(W_\infty)\, \frac{G(x)}{G(0)},$$

for x even (and equal to 0 if x is odd). Here, G is the restriction to even sites of the Green function of the simple random walk,

$$G(x) = P_{0,x}^{\otimes 2}(N_{0,\infty}),$$

and estimates on its decay are extensively used. With the estimates (3.50), (3.51) and (3.52), we conclude that

$$\mathbb{P}\big[(s_n^2 - \sigma_1^2 \mathbb{E} W_n^2)^2\big] = \mathbb{P}[s_n^4] - 2\sigma_1^2 \mathbb{P}[s_n^2 W_n^2] + \sigma_1^4 \mathbb{P}[W_n^4]$$

$$\longrightarrow 0.$$

that is, $s_n^2 - \sigma_1^2 W_n^2 \to 0$ in L^2. As $W_n^2 \to W_\infty^2$ in L^2, it follows that $s_n^2 \to \sigma_1^2 W_\infty^2$ in L^2. We thus obtain (3.48).

To show Lindeberg's condition (3.49) is routine. We check that, for any $q > 1$, when $|\beta| > 0$ is small enough, we have

$$\mathbb{P}[D_{k+1}^4] = O(k^{-d/q}), \qquad k \geq 1.$$

Then, for small β, it holds $n^{(d-2)} \sum_{k \geq n} \mathbb{P}[D_{k+1}^4] \to 0$, yielding (3.49). $\qquad\square$

Open Problem 3.10

(i) *What is the rate of convergence at WD but outside L^2 region? Is the rate of convergence still $n^{-(d-2)/4}$? Do we have stable limits?*

(ii) *Iterated logarithm law for small β: Find a sequence $u_n \to 0$ such that*

$$\limsup_n u_n n^{(d-2)/4}(W - W_n) W_n^{-2} = 1 \quad a.s.$$

Chapter 4
Lattice Versus Tree

In this chapter we deal with polymer models on different oriented graphs and compare them with the lattice case. As revealed by Derrida and Spohn in [94], many interesting questions can be answered on the regular tree. Later on, refined tree-like structures including Derrida's m-tree has been introduced [82], yielding further comparisons. There, correlations are simpler compared to the lattice case, since the medium along two paths becomes independent as soon as they visit different sites. In the sense of simplifying the correlation structure, these models play the role of mean-field models.

We will consider the corresponding partition functions $Z_n^{tr}(\omega; \beta), Z_n^{m-tr}(\omega; \beta)$ and sometimes $Z_n(\omega; \beta) = Z_n^{pol}(\omega; \beta)$. In this section we compare the different models, discussing their similarities as well as their differences. We will use the convention: no superscript means "pol", which refers to the lattice case. In particular, p^{pol} is what we want to talk about!

4.1 A Mean Field Approximation

Rooted **regular tree** with branching number $b = 2d$ is the infinite tree where each vertex has the same number $b + 1$ of neighbors except for one which has only b of neighbors. This particular vertex is called the root or origin of the tree. This lattice is usually called the Bethe lattice with coordination number $b + 1$ in the physics literature. We denote this tree by $\mathbb{T}_b$. It can be encoded by words of lengths $0, 1, \ldots$ in the alphabet $\{1, 2, \ldots, b\}$,

$$\mathbb{T}_b = \emptyset \cup \{1, 2, \ldots, b\} \cup \{1, 2, \ldots, b\}^2 \cdots \cup \{1, 2, \ldots, b\}^n \cup \ldots$$

© Springer International Publishing AG 2017
F. Comets, *Directed Polymers in Random Environments*, Lecture Notes
in Mathematics 2175, DOI 10.1007/978-3-319-50487-2_4

A word $\mathbf{u}$ of length n, or equivalently, a vertex in the tree at the nth generation, represents a path of duration n and the letters are the successive jumps. The environment is now an i.i.d. family $(\omega(\mathbf{u}); \mathbf{u} \in \mathbb{T}_b)$, and

$$H_n^{\text{tr}}(S) = \sum_{t=1}^{n} \omega(\mathbf{u}_t) \qquad \text{with} \quad \mathbf{u}_t = (S_1 - S_0, S_2 - S_1, \ldots, S_t - S_{t-1}) .$$

Hence,

$$\text{Cov}\big(H_n^{\text{tr}}(S), H_n^{\text{tr}}(\tilde{S})\big) = \big|S \wedge \tilde{S}\big| \times \text{Var}(\omega(\mathbf{u})) = \text{Var}(\omega(\mathbf{u})) \times \sum_{t=1}^{n} \prod_{r \in [1,t]} \mathbf{1}_{S_r = \tilde{S}_r} \qquad (4.1)$$

is proportional to the length of the most recent common ancestor $S \wedge \tilde{S}$ of the two paths.

Remark 4.1 (Connection with Branching Random Walks) Consider the branching random walk (BRW) where

- each individual has exactly b direct children,
- each of them performing independently a random jump with the same law as $\omega(0, 0)$.

Each member of the population in the nth generation is encoded by a node $\mathbf{u}$ of length n, and its position is $H_n^{\text{tr}}(\mathbf{u})$ defined above. Then, the renormalized partition function

$$W_n^{\text{tr}} = Z_n^{\text{tr}} e^{-n\lambda}$$

is the so-called Biggins' martingale for the BRW [37]. See [212] for a recent review.

The polymer model on **Derrida's m-tree** was introduced in [82]. It falls into the multiplicative cascades formalism. Multiplicative cascades (or multiplicative chaos) have been introduced by Mandelbrot [168, 169] as a statistical theory of turbulence. Reference [148] makes a systematic study of random measures on the unit interval obtained by random iterated multiplications. In its geometric form the model exhibits a complex multifractal structure. Collet and Koukiou [66] made an early investigation of the model using the thermodynamic formalism. The idea was taken up by Franchi [111]. The free energy of the m-tree can be computed as a generalized multiplicative cascades, see [166], see (4.4) below.

Let us fix an integer $m \geq 1$ and define L_m to be the set of points visited by the simple random walk at time m:

$$L_m \overset{def}{=} \{x \in \mathbb{Z}^d; P(S_m = x) > 0\}. \qquad (4.2)$$

We use a model of generalized multiplicative cascades on a tree. For an overview of results, we refer to [166]. Consider the tree

$$U = \bigcup_{k \in \mathbb{N}} L_m^k$$

be the set of all finite sequences $u = u_1 \ldots u_k$ of elements in L_m. Let q_m be a non degenerate probability distribution on $(\mathbb{R}_+^*)^{L_m}$. Consider a probability space with probability measure denoted by P (and expectation E), and random variables $(A_u)_{u \in U}$ defined on this space, such that the random vectors $(A_{ux}; x \in L_m)_{u \in U}$ form an i.i.d. sequence with common distribution $q_m = \text{Law}(W_m(x); x \in L_m)$. We set the root variable $A_\varnothing$ constant and equal to 1. Then,

$$E\left(\sum_{u \in L_m} A_u\right) = 1.$$

Consider the processes $(Z_{km}^{m-\text{tr}})_{k \in \mathbb{N}}$ and $(W_{km}^{m-\text{tr}})_{k \in \mathbb{N}}$ defined by

$$W_{km}^{m-\text{tr}} = \sum_{u_1, \ldots, u_k \in L_m} A_{u_1} A_{u_1 u_2} \ldots A_{u_1 \ldots u_k} \tag{4.3}$$

for $k \geq 0$. See Fig. 4.1.

This is a non negative martingale so the limit $W_\infty^{m-\text{tr}} = \lim_{k \to \infty} W_{km}^{m-\text{tr}}$ exists, and the associated free energy has a limit

$$p^{m-\text{tr}}(\beta) = \lim_{k \to \infty} \frac{1}{k} \ln Z_{mk}^{m-\text{tr}}.$$

For simplicity we will only consider paths of length $n \in m\mathbb{N}$, and we will not give full details. The polymer on the m-tree can be represented on the lattice by an Hamiltonian with covariance function given, for two "indices" $S, \tilde{S}$, by

$$\text{Cov}_\mathbb{P}\left(H_{km}^{m-\text{tr}}(S), H_{km}^{m-\text{tr}}(\tilde{S})\right) = \text{Var}(\omega(\mathbf{u})) \times \sum_{t=1}^{km} \mathbf{1}_{S_t = \tilde{S}_t} \prod_{r \in m\mathbb{N} \cap [1,t]} \mathbf{1}_{S_r = \tilde{S}_r} \tag{4.4}$$

It is known (e.g. [166]) that

$$p^{m-\text{tr}}(\beta) = \inf_{\theta \in]0,1]} \frac{1}{\theta} \ln(E \sum_{x \in L_m} Z_m(x)^\theta) \leq m\lambda(\beta). \tag{4.5}$$

Note that we recover the standard polymer model on the tree when $m = 1$.

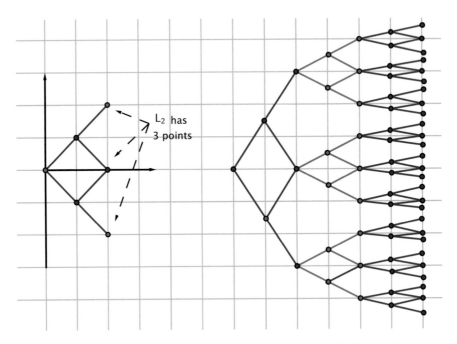

Fig. 4.1 On the *right*, m-tree with $m = 2$ and $k = 3$ generations ($n = 6$). On the *left*, one step (generation) of the tree

The free energy of the lattice polymer relates to one of the m-tree. By Theorems 2.1 and 3.6 in [72], and Theorem 8.1 in [167], we have:

Proposition 4.1 *The m-tree polymer is a consistent approximation of the lattice polymer* ($p = p^{pol}$):

$$p(\beta) = \inf_{m \geq 1} \frac{1}{m} p^{m-tr}(\beta). \tag{4.6}$$

The equality in (4.6) is improving the improved version (2.16) of annealed bound, which corresponds to $m = 1$. Let us mention a characterization (M. Nakashima, 2016, private communication) of the low temperature region of the lattice polymer.

Proposition 4.2 *The following are equivalent* ($p = p^{pol}$):

(i) $p(\beta) < \lambda(\beta)$;
(ii) $\exists m \geq 1 : \mathbb{P}[\sum_x W_m(x) \ln W_m(x)] \geq 0$;
(iii) $\exists m \geq 1 : \mathbb{P}[\sum_x W_m(x) \ln W_m(x)] > 0$.

Proof Statement (i) follows from (iii) by [72, Lemma 2.2].

Conversely, by (4.6), Property (i) implies that there exists $m \geq 1$ such that $p^{m-tr}(\beta) < \lambda(\beta)$. From [72, Lemma 2.2] again, it follows that $\mathbb{P}[\sum_x W_m(x) \ln W_m(x)] > 0$.

Finally, (ii) implies (iii) by a doubling procedure [72, Remark 3.5]. □

Proposition 4.2 gives a characterization of the low temperature region. We mention another one from [137] in the case of Gaussian environment:

$$p(\beta) < \lambda(\beta) \iff \lim_{n \to \infty} n^{-1} \mathbb{P}[W_n \ln W_n] > 0,$$

where the limit is shown to exist and is equal to $\inf_n n^{-1} \mathbb{P}[W_n \ln W_n]$ by subadditivity.

4.2 Majorizing Lattice Polymers by m-Tree Polymers

Here we fix a temperature β, which will be used for each model. We have seen in Proposition 4.1 that

$$\mathbb{P} \ln Z_{mk}^{\text{pol}} \leq \mathbb{P} \ln Z_{mk}^{m-\text{tr}} \leq \mathbb{P} \ln Z_{mk}^{\text{tr}}. \tag{4.7}$$

In the Gaussian case, it is a consequence of Slepian's lemma (Theorem A.3): Indeed, $H_n^{\text{pol}}, H_n^{m-\text{tr}}, H_n^{\text{tr}}$ are Gaussian processes indexed by paths, with

$$\text{Cov}(H_n^{\text{tr}}(S), H_n^{\text{tr}}(\tilde{S})) \leq \text{Cov}(H_n^{m-\text{tr}}(S), H_n^{m-\text{tr}}(\tilde{S})) \leq \text{Cov}(H_n^{\text{pol}}(S), H_n^{\text{pol}}(\tilde{S})),$$

with equality if $S = \tilde{S}$, and the free energy has non-positive, mixed second derivative.

It is tempting to look for a stochastic order for which the r.v.'s $\ln Z_{mk}$ themselves are ordered. This is not true for the usual stochastic order for real r.v.'s ($X \leq_{\text{st}} Y$ iff $\mathbb{P}[f(X)] \leq \mathbb{P}[f(Y)]$ for all increasing f), even in the Gaussian case. However, it is true for general distribution of the disorder, in the Laplace transform order.

Definition 4.1 For two positive random variables X, Y, we say that X is smaller than Y in the Laplace transform order, and we write $X \leq_{\text{Lt}} Y$, iff

$$\forall u \geq 0, \quad \mathbb{P}[-e^{-uX}] \leq \mathbb{P}[-e^{-uY}].$$

It is straightforward to see that

$$X \leq_{\text{Lt}} Y \iff (X/\varepsilon) \leq_{\text{st}} (Y/\varepsilon')$$

where $\varepsilon, \varepsilon'$ are Exponential(1) r.v.'s with ε independent of X, ε' independent of Y.

Theorem 4.1 ([188])

$$Z_{mk}^{\text{pol}} \leq_{\text{Lt}} Z_{mk}^{m-\text{tr}} \leq_{\text{Lt}} Z_{mk}^{\text{tr}},$$

and similarly with point-to-point partition functions.

It is a general fact (see [188] for details) that (4.7) follows from this, as well as

$$\mathbb{P}[(Z_{mk}^{\text{pol}})^\alpha] \leq \mathbb{P}[(Z_{mk}^{m-\text{tr}})^\alpha] \leq \mathbb{P}[(Z_{mk}^{\text{tr}})^\alpha], \quad \alpha \in (0,1),$$

and, since the three partition functions have same expectation, that

$$\mathbb{P}[Z_{mk}^{\text{pol}} \ln Z_{mk}^{\text{pol}}] \geq \mathbb{P}[Z_{mk}^{m-\text{tr}} \ln Z_{mk}^{m-\text{tr}}] \geq \mathbb{P}[Z_{mk}^{\text{tr}} \ln Z_{mk}^{\text{tr}}].$$

4.3 Phase Diagram on the Tree

In this Section, we compute explicitly the critical temperature and the value of the free energy on the b-ary tree with $b = 2d$.

Similar to the definition of β_1 in Proposition 2.2, we define

$$\beta_c^{\text{tr}} = \begin{cases} \text{the unique root } \beta \in (0,+\infty) \text{ of } \beta\lambda'(\beta) - \lambda(\beta) = \ln(2d) \text{ , if it exists,} \\ +\infty, \qquad\qquad\qquad\qquad\qquad\qquad\qquad\qquad\qquad\qquad \text{otherwise.} \end{cases} \tag{4.8}$$

If $\beta_c^{\text{tr}} < \infty$, then $\lambda'(\beta_c^{\text{tr}}) = a$ is such that $a > \mathbb{P}(\omega(u))$ and $\lambda^*(a) = \ln b$. In particular,

$$\lambda'(\beta_c^{\text{tr}}) = \sup\{a : \lambda^*(a) \leq \ln b\} = \inf\left\{\frac{\lambda(\beta) + \ln b}{\beta}; \beta > 0\right\} \tag{4.9}$$

for $\beta_c^{\text{tr}} < \infty$ and similar for $\lim_{b\to\infty} \lambda'(\beta)$ when $\beta_c^{\text{tr}} = \infty$. The last term already appeared with $b = 2d$ in (2.19) and the lines below it. We build on this analogy.

Observe that the annealed bound applies, as well as its improved version (2.16). An interesting fact is that, the free energy on the tree is exactly given by this upper bound, which is the content of (4.10).

Theorem 4.2 *The free energy $n^{-1}\ln Z_n^{\text{tr}}(\omega;\beta)$ converges almost surely and in L^p-norm $(1 \leq p < \infty)$ to $p^{\text{tr}}(\beta)$, with*

$$p^{\text{tr}}(\beta) = \begin{cases} \lambda(\beta) & \text{if } \beta \leq \beta_c^{\text{tr}}, \\ \beta\lambda'(\beta_c^{\text{tr}}) - \ln(2d) & \text{if } \beta > \beta_c^{\text{tr}}. \end{cases} \tag{4.10}$$

Moreover, letting $\bar{\beta}_c^{\text{tr}} = \inf\{\beta \geq 0 : W_\infty^{\text{tr}} = 0\}$, we have

$$\beta_c^{\text{tr}} = \bar{\beta}_c^{\text{tr}}. \tag{4.11}$$

The function in (4.10) is equal to the improved annealed bound given in (2.16), cf. Fig. 2.2.

Remark 4.2 (The Phase Transition on the Tree is Second Order) The limit p^{tr} is $\mathscr{C}^\infty$ except at the critical value β_c^{tr}. At β_c^{tr} it is once continuously differentiable in β, but not $\mathscr{C}^2$. The phase transition at β_c^{tr} of is second order, meaning that there is a jump in the second derivative at the critical value.

Proof First, observe that we can use the martingale approach developed in Sect. 3.1: The sequence $W_n^{\text{tr}} = Z_n^{\text{tr}}(\omega; \beta)e^{-n\lambda(\beta)}$ from (3.1) is still a positive martingale, it converges a.s. to some non-negative W_∞^{tr}, which is either a.s. equal to 0, or a.s. positive. In the last case, we have $p^{\text{tr}}(\beta) = \lambda(\beta)$.

In order to study the positivity of the limit, we will crucially use an inequality from Kahane and Peyrière [148]. By subadditivity of the power $\alpha \in (0, 1)$,

$$(\sum_{i=1}^{b} x_i)^\alpha \le \sum_{i=1}^{b} x_i^\alpha, \qquad b \ge 1, x_i \ge 0.$$

The following lemma gives an estimate in the reverse direction.

Lemma 4.1 ([148]) *There exists $\alpha_0 \in (0, 1)$ such that, for all $\alpha \in (\alpha_0, 1)$, for all $b \ge 2$,*

$$(\sum_{i=1}^{b} x_i)^\alpha \ge \sum_{i=1}^{b} x_i^\alpha - 2(1-\alpha) \sum_{1 \le i < j \le b} (x_i x_j)^{\alpha/2}, \qquad x_i > 0, i = 1, \ldots, b. \quad (4.12)$$

We will prove the lemma later. We first complete the proof of Theorem 4.2 in three steps.

Step 1: for $\beta < \beta_c^{\text{tr}}$, $W_\infty^{\text{tr}} > 0$ almost surely.
Putting $W_{n-1,i}^{\text{tr}} = W_{n-1}^{\text{tr}} \circ \theta_i$ with θ_i the shift, we have by the property analogous to (2.3) for the tree,

$$W_n^{\text{tr}} = b^{-1} \sum_{i=1}^{b} \exp\{\beta\omega(i) - \lambda(\beta)\} \times W_{n-1,i}^{\text{tr}}.$$

From (4.12) and by the independence of the variables in the right-hand side, we obtain

$$
\begin{aligned}
b^\alpha \mathbb{P}((W_n^{\text{tr}})^\alpha) &\ge b\mathbb{P}(e^{\alpha\beta\omega(i)-\alpha\lambda(\beta)})\mathbb{P}((W_{n-1}^{\text{tr}})^\alpha) \\
&\quad - b(b-1)(1-\alpha)\mathbb{P}(e^{[\alpha\beta\omega(i)-\alpha\lambda(\beta)]/2})^2\mathbb{P}((W_{n-1}^{\text{tr}})^{\alpha/2})^2 \\
&= be^{\lambda(\alpha\beta)-\alpha\lambda(\beta)}\mathbb{P}((W_{n-1}^{\text{tr}})^\alpha) \\
&\quad - b(b-1)(1-\alpha)e^{2\lambda(\alpha\beta/2)-\alpha\lambda(\beta)}\mathbb{P}((W_{n-1}^{\text{tr}})^{\alpha/2})^2.
\end{aligned}
$$

By Jensen's inequality for the concave function $x \mapsto x^\alpha$ on $\mathbb{R}_+^*$, we have $\mathbb{P}((W_n^{\text{tr}})^\alpha) \leq \mathbb{P}((W_{n-1}^{\text{tr}})^\alpha)$, and by convexity we have $2\lambda(r/2) \leq \lambda(r)$. Hence,

$$\left(e^{(1-\alpha)\ln b - \alpha\lambda(\beta) + \lambda(\alpha\beta)} - 1\right)\mathbb{P}((W_n^{\text{tr}})^\alpha) \leq (1-\alpha)b^{1-\alpha}(b-1)\mathbb{P}((W_{n-1}^{\text{tr}})^{\alpha/2})^2 \; .$$

Dividing both sides by $1 - \alpha$ and letting $\alpha \nearrow 1$, we obtain

$$\ln b - \beta\lambda'(\beta) + \lambda(\beta) \leq (b-1)\mathbb{P}((W_{n-1}^{\text{tr}})^{1/2})^2 \; .$$

Since $\mathbb{P}[((W_n^{\text{tr}})^{1/2})^2] = 1$, the sequence $(W_n^{1/2})$ is bounded in L^2, hence it is uniformly integrable, which, together with almost sure convergence, implies that

$$\mathbb{P}((W_{n-1}^{\text{tr}})^{1/2}) \to \mathbb{P}((W_\infty^{\text{tr}})^{1/2}), \qquad n \to \infty.$$

Finally, if $\beta < \beta_c^{\text{tr}}$,

$$0 < \ln b - \beta\lambda'(\beta) + \lambda(\beta) \leq (b-1)\mathbb{P}((W_\infty^{\text{tr}})^{1/2})^2 \; ,$$

which implies $W_\infty^{\text{tr}} \neq 0$, and by the 0–1 law, $W_\infty^{\text{tr}} > 0$ a.s. We have proved that (4.10) holds for $\beta < \beta_c^{\text{tr}}$, and also for $\beta \leq \beta_c^{\text{tr}}$ by continuity.

Step 2: For $\beta \geq \beta_c^{\text{tr}}$, $\quad \liminf_n p_n^{\text{tr}}(\omega; \beta) \geq \beta\lambda'(\beta_c^{\text{tr}}) - \ln b$.

Indeed, $p_n^{\text{tr}}(\omega; \beta)$ is convex in β, hence, for $\beta \geq \beta_c^{\text{tr}}$ we have a.s.,

$$\begin{aligned}
\liminf_n \frac{d}{d\beta}p_n^{\text{tr}}(\omega; \beta) &\geq \sup_{\beta' < \beta_c^{\text{tr}}} \lim_n \frac{d}{d\beta}p_n^{\text{tr}}(\omega; \beta') \\
&= \sup_{\beta' < \beta_c^{\text{tr}}} \lambda'(\beta') \\
&= \lambda'(\beta_c^{\text{tr}})
\end{aligned}$$

where the equalities come from Step 1. By integration from β_c^{tr} to β and using that $p^{\text{tr}}(\beta_c^{\text{tr}}) = \lambda(\beta_c^{\text{tr}})$, we get the claim.

Step 3: To get the reverse inequality, we let

$$H_n^* = \max\{H_n^{\text{tr}}(\omega); \omega \text{ path of length } n\},$$

and we start to show that

$$\limsup_{n\to\infty} n^{-1}H_n^* \leq \lambda'(\beta_c^{\text{tr}}) \; , \tag{4.13}$$

with the convention $\lambda'(\beta_c^{\text{tr}}) = \lim_{\beta\to\infty}\lambda'(\beta) = s$ when $\beta_c^{\text{tr}} = \infty$. The other case being trivial, we assume β_c^{tr} finite, and we fix $a > \lambda'(\beta_c^{\text{tr}})$. By the union bound and Cramer's theorem

$$\mathbb{P}(H_n^* \geq an) \leq b^n \exp\{-n\lambda^*(a) + o(n)\} \; ,$$

which vanishes exponentially fast since $\lambda^*(a) > \lambda^*[\lambda'(\beta_c^{tr})] = \ln b$. The claim (4.13) follows from an application of Borel-Cantelli lemma.

By convexity, for $\beta \geq \beta_c^{tr}$,

$$p_n^{tr}(\omega; \beta) \leq p_n^{tr}(\omega; \beta_c^{tr}) + (\beta - \beta_c^{tr})\frac{d}{d\beta}p_n^{tr}(\omega; \beta)$$

Since the derivative above is $n^{-1}P_n^{\beta,\omega}(H_n^{tr}) \leq n^{-1}H_n^*$, we get

$$p^{tr}(\beta) \leq p^{tr}(\beta_c^{tr}) + (\beta - \beta_c^{tr})\lambda'(\beta_c^{tr}) = \beta\lambda'(\beta_c^{tr}) - \ln b$$

$\mathbb{P}$-a.s. This proves the almost sure convergence in the theorem. To get L^p-convergence, one can derive a concentration inequality and proceed as in the proof of Theorem 2.1.

Step 3: (4.11) follows from Step 1.

$\square$

We now prove Lemma 4.1.

Proof Following [148], we prove the lemma by induction, starting with the case $b = 2$. Let

$$f(t) = e^{\alpha t} + e^{-\alpha t} - (e^t + e^{-t})^\alpha, \quad h(t) = 2\frac{e^{(1-\alpha)t} - e^{-(1-\alpha)t}}{e^t - e^{-t}},$$

and $C_\alpha = \sup_{t \in \mathbb{R}} f(t) = \sup_{t>0} f(t)$. Dividing both hands of (4.12) for $b = 2$ by $(x_1 x_2)^{\alpha/2}$, and letting $t = \ln(x_1/x_2)$ in (4.12), we see that all what we need is to prove that

$$C_\alpha \overset{\text{def.}}{=} \sup_{t \in \mathbb{R}} f(t) \leq 2(1 - \alpha)$$

for all α as in the lemma. The function f is non negative by subadditivity of $r \mapsto r^\alpha$, it is equal to 0 at 0, and vanishes at infinity. It is complicated to study its maxima, but the point is that $f(t) = h(t)$ at each point $t > 0$ with $f'(t) = 0$: to check this, one can write

$$\begin{aligned} f'(t) &= \alpha(e^{\alpha t} - e^{-\alpha t}) - \alpha(e^t - e^{-t})(e^t + e^{-t})^{\alpha-1} \\ &= \frac{\alpha}{e^t + e^{-t}}\left[(e^{\alpha t} - e^{-\alpha t})(e^t + e^{-t}) - (e^t - e^{-t})(e^t + e^{-t})^\alpha\right] \\ &= \frac{\alpha}{e^t + e^{-t}}\left[f(t)(e^t - e^{-t}) - 2(e^{(1-\alpha)t} - e^{-(1-\alpha)t})\right] \\ &= \frac{\alpha(e^t - e^{-t})}{e^t + e^{-t}}[f(t) - h(t)]. \end{aligned}$$

Then, for all extrema t of f it holds $f(t) \leq h(t)$, and therefore

$$C_\alpha \leq \sup_{t>0} h(t) \,.$$

Next we define, for all $t > 0$, the function $g = g_t$ on $\mathbb{R}^+$,

$$g(a) = e^{at} - e^{-at} - a(e^t - e^{-t}) \,,$$

and we observe that g vanishes at 0 and has a negative derivative for small a. For this last point, we can expand g' in Taylor series

$$g'(a) = \sum_{k \geq 0} \left((2k+1)a^{2k} - 1 \right) \frac{t^{2k+1}}{(2k+1)!},$$

which is negative for all t and all $a < 1/\sqrt{3}$. Finally, we obtain that $C_\alpha \leq 2(1-\alpha)$ for all $\alpha \in (\alpha_0, 1)$ with $\alpha_0 = 1 - 1/\sqrt{3}$, and the inequality (4.12) is proved for $b = 2$.

The case of a general b is obtained by induction on b. From the case $b = 2$,

$$\left(\sum_{i=1}^{b} x_i \right)^\alpha \geq x_1^\alpha + \left(\sum_{i=2}^{b} x_i \right)^\alpha - 2(1-\alpha)x_1^{\alpha/2} \left(\sum_{i=2}^{b} x_i \right)^{\alpha/2}$$

$$\geq x_1^\alpha + \left(\sum_{i=2}^{b} x_i \right)^\alpha - 2(1-\alpha) \sum_{2 \leq j \leq b} (x_1 x_j)^{\alpha/2} \,,$$

where the last inequality follows from the subadditivity of the function $x^{\alpha/2}$. Using now the induction assumption for $b - 1$, we obtain the desired inequality for b. □

It is not difficult to derive an important consequence about the ground state energy, i.e., the maximal energy over the configuration space. Recall the notation $s = \sup \mathrm{supp}[q]$ from Corollary 2.1.

Corollary 4.1 *The following convergence holds* $\mathbb{P}$*-a.s. and in* $L^p, p < \infty$,

$$n^{-1} \max\{H_n^{\mathrm{tr}}(\mathbf{x}); \mathbf{x} \text{ path of length } n\} \longrightarrow \lambda'(\beta_c^{\mathrm{tr}}) \,,$$

with the convention $\lambda'(+\infty) = \lim_{\beta \to +\infty} \lambda'(\beta) = s$.

Remark 4.3 (Linear Part of the Free Energy on the Tree) The picture beyond (4.10) is that, on the tree and for β smaller than the critical value, the polymer measure is carried out by a "large" set of paths with energy $H_n^{\mathrm{tr}} \simeq n\lambda'(\beta)$. This set changes when β is increased. At $\beta = \beta_c^{\mathrm{tr}}$ the maximal energy $\max_{\mathbf{x}} H_n^{\mathrm{tr}} \simeq n\lambda'(\beta_c^{\mathrm{tr}})$ has been reached (cf. Corollary 4.1), and from that temperature the measure concentrates on the "small" set of (quasi)-optimal paths.[1] When β increases beyond β_c^{tr}, no significant changes occur at a macroscopic level, since for all $\beta, \beta' > \beta_c^{tr}$, $\mathcal{H}(P_n^{\beta,\omega}|P_n^{\beta',\omega}) = o(n)$ alsmost surely. This is responsible for the linear part of the free energy.

The model on $\mathbb{Z}^d$ is very different in this regard: Anticipating on Theorem 4.3 and its Corollary 4.2 below, we see that any change in β creates macroscopic changes in the polymer measure. The function p^{pol} is strictly convex, the asymptotic energy per monomer $\lim\sup_n n^{-1}P_n^{\beta,\omega}(H_n^{pol})$ is strictly increasing in β.

4.4 Free Energy is Strictly Convex for the Polymer on the Lattice

In this Section, we only consider the model on the lattice, so we drop the indication "pol" in the notations.

It is easy to see, by checking the second derivative, that p_n is a strictly convex function of β, for a.e. environment. Taking the limit $n \to \infty$ preserves convexity, but strict convexity may be lost for some values of β. We have seen in Theorem 4.2 that this is the case for polymer *on the tree*, where the *free energy is linear* on the whole interval $[\beta_c^{tr}, \infty[$. However, the situation is *drastically different* on $\mathbb{Z}^d$.

In the rest of the chapter, we consider the model on $\mathbb{Z}^d$. For the sake of simplicity, we make in this section the **extra assumption**[2] that $\omega(t, x)$ is a Bernoulli variable with parameter $q \in (0, 1)$,

$$\mathbb{P}(\omega(t, x) = 1) = q, \quad \mathbb{P}(\omega(t, x) = 0) = 1 - q. \tag{4.14}$$

The main result in this section is taken from [80]:

Theorem 4.3 *Assume Bernoulli environment (4.14). Then, the free energy $p(\beta)$ of the polymer model on $\mathbb{Z}^d$ is a strictly convex function on $\mathbb{R}$.*

The main consequence of the strict convexity of the free energy is that the polymer measures $P_n^{\beta,\omega}$ look quite different for different ω's. Heuristically, strict convexity means that each possible value for the average energy $\lim_n \mathbb{P}P_n^{\beta,\omega}(H_n/n)$—if the limit exists (but this will be the case for all but countably many β's)—corresponds to a *unique* value of the parameter β. It prevents the average energy to be constant over some range of temperatures. Recall the definition (3.41) of the relative entropy

---

[1] Here, "large" means exponential in n, and "small" means subexponential—the bound on the entropy has been reached.

[2] The proof below can be directly adapted to the case of a bounded, but otherwise general, environment η, by dividing the range of $\eta(t, x)$ into three intervals, to take into account the variability of η. We will not further address this issue.

$\mathcal{H}(\nu|\mu) \geq 0$, which measures the discrepancy between the probability measures ν and μ.

Corollary 4.2 *Assume (4.14). For all finite r, there exists $C > 0$ such that we have for all $\beta_1, \beta_2 \in [0, r]$,*

$$\mathbb{P}[\mathcal{H}(P_n^{\beta_1,\omega}|P_n^{\beta_2,\omega}) + \mathcal{H}(P_n^{\beta_2,\omega}|P_n^{\beta_1,\omega})] \geq nC(\beta_1 - \beta_2)^2$$

The inequality means that the measures $P_n^{\beta,\omega}$ changes significantly at a macroscopic scale as the temperature is varied. Indeed, entropy is an extensive quantity, and the magnitude of the entropy between two "really different" measures is of order of the length n of paths. This situation holds because the walk takes place on the lattice $\mathbb{Z}^d$. We have seen in Theorem 4.2 that the situation is quite different, at low temperature, when the walk takes place on a tree.

We now state the main ingredient of the proof of the theorem, a variance estimate analogous to that for Gibbs random fields in [100].

Lemma 4.2 *Assume Bernoulli environment (4.14). For any compact set $K \subset \mathbb{R}$, there exists a positive constant $C = C_K$ such that*

$$\mathbb{P}\frac{d^2}{d\beta^2} \ln Z_n(\omega; \beta) \geq Cn, \qquad \beta \in K.$$

Note Theorem 4.3 is from [80], and the idea of bounding below with the conditional variance as we will see in the proof of the lemma has been borrowed to the theory of Gibbs fields, where it goes back at least to [100]. The consequence of strict convexity to proper parametrization of the Gibbs state was already stated in [124].

Proof (Proof of Corollary 4.2) By definition of the entropy, we have

$$\mathcal{H}(P_n^{\beta_1,\omega}|P_n^{\beta_2,\omega}) = P_n^{\beta_1,\omega}\left[\ln \frac{dP_n^{\beta_1,\omega}}{dP_n^{\beta_2,\omega}}\right]$$

$$= P_n^{\beta_1,\omega}\left[(\beta_1 - \beta_2)H_n - \ln \frac{Z_n(\omega; \beta_1)}{Z_n(\omega; \beta_2)}\right]$$

$$= (\beta_1 - \beta_2)P_n^{\beta_1,\omega}[H_n] - \ln \frac{Z_n(\omega; \beta_1)}{Z_n(\omega; \beta_2)}.$$

Therefore, the partition functions cancel by symmetrization,

$$\mathcal{H}(P_n^{\beta_1,\omega}|P_n^{\beta_2,\omega}) + \mathcal{H}(P_n^{\beta_2,\omega}|P_n^{\beta_1,\omega}) = (\beta_1 - \beta_2)[P_n^{\beta_1,\omega}(H_n) - P_n^{\beta_2,\omega}(H_n)]$$

$$= n\left[\frac{d}{d\beta}p_n(\omega, \beta_1) - \frac{d}{d\beta}p_n(\omega, \beta_2)\right](\beta_1 - \beta_2).$$

Now, by Lemma 4.2, for all bounded interval $I \subset \mathbb{R}^+$, there exists $C > 0$ depending on I such that $\mathbb{P} \frac{d^2}{d\beta^2} p_n(\omega; \beta) \geq 2C$ for all $\beta \in I$. Thus,

$$[\mathbb{P} \frac{d}{d\beta} p_n(\omega, \beta_1) - \mathbb{P} \frac{d}{d\beta} p_n(\omega, \beta_2)][\beta_1 - \beta_2] \geq C[\beta_1 - \beta_2]^2 .$$

Then, the desired estimate follows by integrating with respect to $\mathbb{P}$ the last equality.

$\square$

Assuming Lemma 4.2 for the moment, we complete the

Proof (Proof of Theorem 4.3) It follows from Lemma 4.2 that, for $\beta, \beta' \in K$,

$$p(\beta') \geq p(\beta) + (\beta' - \beta) p'_r(\beta) + \frac{C_K}{2}(\beta' - \beta)^2 , \quad \beta \leq \beta' ,$$

with p'_r the right-derivative, and a similar statement for $\beta' \leq \beta$. Indeed, this inequality holds for $(1/n)\mathbb{P} \ln Z_n(\omega; \beta)$ instead of p, and we can pass to the limit $n \to \infty$. This yields the strict convexity of p.

$\square$

Proof (Proof of Lemma 4.2) The polymer measure $P_n^{\beta,\omega}$ is Markovian (but time-inhomogeneous), and

$$\frac{d^2}{d\beta^2} \ln Z_n(\omega; \beta) = \mathrm{Var}_{P_n^{\beta,\omega}}(H_n).$$

Define

$$\mathscr{I}(x, y) = \{z \in \mathbb{Z}^d : \|x - z\|_1 = \|z - y\|_1 = 1\}, \quad x, y \in \mathbb{Z}^d \tag{4.15}$$

the set of lattice points which are next to both x and y. The set $\mathscr{I}(x, y)$ is empty except if y can be reached in two steps by the simple random walk from x; in this case its cardinality is equal to $2d, 2$ or 1 according to $y = x, \|y - x\|_\infty = 1$ or $\|y - x\|_\infty = 2$. The Markov property implies that, under $P_{2n}^{\beta,\omega}$, $S_1, S_3, \ldots, S_{2n-1}$ are independent conditionally on $S^e := (S_2, S_4, \ldots, S_{2n})$, and the law of S_{2t-1} given S^e only depends on S_{2t-2}, S_{2t}, and has support $\mathscr{I}(S_{2t-2}, S_{2t})$.

From the variance decomposition under conditioning, we have

$$\mathrm{Var}_{P_{2n}^{\beta,\omega}}(H_{2n}) = P_{2n}^{\beta,\omega}[\mathrm{Var}_{P_{2n}^{\beta,\omega}}(H_{2n} \mid S^e)] + \mathrm{Var}_{P_{2n}^{\beta,\omega}}(P_{2n}^{\beta,\omega}[H_{2n} \mid S^e])$$

$$\geq P_{2n}^{\beta,\omega}[\mathrm{Var}_{P_{2n}^{\beta,\omega}}(H_{2n} \mid S^e)]$$

$$= P_{2n}^{\beta,\omega}[\mathrm{Var}_{P_{2n}^{\beta,\omega}}(\sum_{t=1}^{n} \omega(2t - 1, S_{2t-1}) \mid S^e)]$$

$$= \sum_{t=1}^{n} P_{2n}^{\beta,\omega}[\mathrm{Var}_{P_{2n}^{\beta,\omega}}(\omega(2t - 1, S_{2t-1}) \mid S^e)]$$

where $P_{2n}^{\beta,\omega}[\ \cdot\ |\ S^e]$, $\mathrm{Var}(\ \cdot\ |\ S^e)$ denote conditional expectation and conditional variance. To obtain the last equality we used the conditional independence. Define the event

$$M(\omega, t, y, z) = \left\{\mathrm{Card}\{\omega(t, x); x \in \mathscr{I}(y, z)\} = 2\right\}.$$

The reason for introducing $M(\omega, t, y, z)$ is that on this event, a path S conditioned on $S_{t-1} = y$, $S_{t+1} = z$, has the option to pick up a $\omega(t, S_t)$ value that can be either 0 or 1, therefore bringing some amount of randomness. This event plays a key role here. Note for further purpose that

$$\mathbb{P}\Big(M(\omega, t, y, z)\Big) = 1 - \left(q^{\mathrm{Card}\,\mathscr{I}(y,z)} + (1-q)^{\mathrm{Card}\,\mathscr{I}(y,z)}\right) =: \bar{q}(y - z) \qquad (4.16)$$

(recall that q is the parameter of the Bernoulli $\omega(t, x)$). The key observation is, for all $t \le n$ and $\beta \in K$,

$$\mathrm{Var}_{P_{2n}^{\beta,\omega}}(\omega(2t - 1, S_{2t-1}) \mid S^e) \ge C\mathbf{1}\{M(\omega, 2t - 1, S_{2t-2}, S_{2t})\}, \qquad (4.17)$$

where the constant C depends only on K and the dimension d. Indeed, on the event $M(\omega, 2t - 1, S_{2t-2}, S_{2t})$, the variable $\omega(2t - 1, S_{2t-1})$ brings some fluctuation under the conditional law: it takes values 0 and 1 with probability uniformly bounded away from 0 provided β remains in the compact set. Hence,

$$\mathbb{P}\mathrm{Var}_{P_{2n}^{\beta,\omega}}(H_{2n}) \ge C\mathbb{P}\sum_{t=1}^{n} P_{2n}^{\beta,\omega}[M(\omega, 2t - 1, S_{2t-2}, S_{2t})]$$

$$= C\mathbb{P}\sum_{t=1}^{n}\sum_{x,y\in\mathbb{Z}^d} P_{2n}^{\beta,\omega}(S_{2t-2} = x, S_{2t} = y)\mathbf{1}\{M(\omega, 2t - 1, x, y)\}.$$

For $1 \le i \le n$, let $\tilde{\mu}_n^{(i)}$ be the polymer measure in the environment $\tilde{\omega}(t, x) = \omega(t, x)$ if $t \neq i$, $\tilde{\omega}(i, x) = 0$ for all x. Obviously, since ω is bounded,

$$C^- \tilde{\mu}_n^{(i)}(\mathbf{x}) \le P_n^{\beta,\omega}(S = \mathbf{x}) \le C^+ \tilde{\mu}_n^{(i)}(\mathbf{x}), \qquad \forall \mathbf{x} \text{ path of length } n,$$

with positive finite C^-, C^+ not depending on $n, \omega, \beta \in K$. The choice of $\tilde{\mu}_n^{(i)}$ is for having this random measure independent of $\omega(i, \cdot)$, a property which is used to get the equality in the next display. With $C' = CC^-$,

$$\mathbb{P}\mathrm{Var}_{P_{2n}^{\beta,\omega}}(H_{2n})$$

$$\ge C'\mathbb{P}\sum_{t=1}^{n}\sum_{x,y\in\mathbb{Z}^d} \tilde{\mu}_{2n}^{(2t-1)}(S_{2t-2}=x, S_{2t}=y)\mathbf{1}\{M(\omega, 2t - 1, x, y)\}$$

$$\overset{\text{indep.}}{=} C'\mathbb{P}\sum_{t=1}^{n}\sum_{x,y\in\mathbb{Z}^d}\tilde{\mu}_{2n}^{(2t-1)}(S_{2t-2}=x, S_{2t}=y)\mathbb{P}(M(\omega, 2t-1, x, y))$$

$$\geq 2C'q(1-q)\mathbb{P}\sum_{t=1}^{n}\sum_{x,y\in\mathbb{Z}^d}\tilde{\mu}_{2n}^{(2t-1)}(S_{2t-2}=x, S_{2t}=y)\mathbf{1}\{\|x-y\|_\infty \leq 1\}$$

$$\geq C'q(1-q)\mathbb{P}\sum_{t=2}^{2n}\sum_{x,y\in\mathbb{Z}^d}\tilde{\mu}_{2n}^{(2t-1)}(S_{t-2}=x, S_t=y)\mathbf{1}\{\|x-y\|_\infty \leq 1\},$$

since we can repeat the same procedure, but conditioning on the path at odd times. We put $\Delta S_t := S_t - S_{t-1}$ and $C'' = (C'/C^+)q(1-q)$, and we note that

$$\|S_t - S_{t-2}\|_\infty \leq 1 \iff \Delta S_t \neq \Delta S_{t-1}.$$

Then, for all $\beta \in K, \epsilon > 0$,

$$\mathbb{P}\text{Var}_{P_{2n}^{\beta,\omega}}(H_{2n}) \geq C''\mathbb{P}P_{2n}^{\beta,\omega}\Big[\sum_{t=2}^{2n}\mathbf{1}\{\Delta S_t \neq \Delta S_{t-1}\}\Big]$$

$$\geq nC''\epsilon \times \mathbb{P}P_{2n}^{\beta,\omega}(A_{n,\epsilon}), \qquad (4.18)$$

where

$$A_{n,\epsilon} = \Big\{S \in \mathscr{P}_n : \sum_{t=2}^{2n}\mathbf{1}\{\Delta S_t \neq \Delta S_{t-1}\} \geq n\epsilon\Big\}.$$

We now finish the proof by showing that $\lim_{n\to\infty}\mathbb{P}P_{2n}^{\beta,\omega}(A_{n,\epsilon}) = 1$ if $\epsilon > 0$ is small enough. It is easy to see that the complement

$$A_{n,\epsilon}^c = \Big\{\sum_{t=2}^{2n}\mathbf{1}\{\Delta S_t = \Delta S_{t-1}\} > n(2-\epsilon)\Big\}$$

of this set has cardinality smaller than $\exp\{2n\delta(\epsilon)\}$, with $\delta(\epsilon) \searrow 0$ as $\epsilon \searrow 0$. We bound

$$\mathbb{P}(\max\{H_{2n}(S); S \in A_{n,\epsilon}^c\} \geq 2n\rho) \leq e^{2n\delta(\epsilon)} \times \text{Prob}(\mathscr{B}(2n, q) \geq 2n\rho),$$

with $\mathscr{B}(2n, q)$ a binomial random variable. It follows that there exists some $\rho(\epsilon)$ with $\rho(\epsilon) \searrow q$ as $\epsilon \searrow 0$ such that the left-hand side is less than $\exp\{-2n\delta(\epsilon)^{1/2}\}$. Moreover, we know from Theorem 2.1 that for positive ϵ,

$$\delta_\epsilon'(n) := \mathbb{P}(|\ln Z_{2n,\beta}^\omega - 2np(\beta)| \geq 2n\epsilon) \longrightarrow 0$$

as $n \to \infty$. Now, for all ω such that $\max\{H_{2n}(S); S \in A_{n,\epsilon}^c\} \le 2n\rho(\epsilon)$ and such that $|\ln Z_{2n}(\omega; \beta) - 2np(\beta)| \le 2n\epsilon$, we have the estimate

$$
\begin{aligned}
P_{2n}^{\beta,\omega}(A_{n,\epsilon}^c) &\le (Z_{2n}(\omega; \beta))^{-1} e^{\beta \max\{H_{2n}(S); S\in A_{n,\epsilon}^c\}} \mathrm{Card}(A_{n,\epsilon}^c)(2d)^{-2n} \\
&\le \exp\{2n[\beta\rho(\epsilon) - p(\beta) - \ln(2d) + \delta(\epsilon) + \epsilon]\} \\
&\le \exp\{2n[p^*(\rho(\epsilon)) + \delta(\epsilon) - \ln(2d) + \epsilon]\}.
\end{aligned}
$$

Here, the Legendre transform is denoted by $*$, $p^*(r) = \sup_{\beta \in \mathbb{R}}\{\beta r - p(\beta)\}$. We will bound

$$
\mathbb{P}P_{2n}^{\beta,\omega}(A_{n,\epsilon}^c) \le e^{-2n\delta(\epsilon)^{1/2}} + \delta_\epsilon'(n) + e^{2n[p^*(\rho(\epsilon))+\delta(\epsilon)-\ln(2d)+\epsilon]}. \tag{4.19}
$$

But, as $\epsilon \searrow 0$,

$$
p^*(\rho(\epsilon)) + \delta(\epsilon) + \epsilon - \ln(2d) \to p^*(q) - \ln(2d) = -\ln(2d) < 0.
$$

By continuity we can choose $\epsilon > 0$ such that $p^*(\rho(\epsilon)) + \delta(\epsilon) \le (-1/2)\ln(2d)$, and

$$
\mathbb{P}P_{2n}^{\beta,\omega}(A_{n,\epsilon}) \to 1 \qquad \text{as } n \to \infty.
$$

Finally, from (4.18) we obtain the desired result for even n. The same computations apply to $P_{2n+1}^{\beta,\omega}$, yielding a similar bound. This concludes the proof of Lemma 4.2.

$\square$

4.5 Conclusions and Related Models

In this chapter, we have defined useful polymer models on the tree (or branching random walks) considered in [94], and discussed both their relations and differences with the model on the lattice. A refined version, the m-tree has been introduced in [82]. All these models majorize the model on the lattice (cf. Theorem 4.1), and their free energy converges to the one on the lattice as $m \to \infty$.

There are fundamental discrepancies between the tree and the lattice, making some asymptotic properties quite different. Many of them follow from this observation: two paths on the tree will never meet again after they separate, though on the lattice, further intersections between paths are always possible. We have already spotted two major differences between the two models.

- The free energy is strictly convex on the lattice, but on the tree it is linear for $\beta \ge \beta_c^{\mathrm{tr}}$, cf. Theorem 4.3. Note that β_c^{tr} may be infinite.
- Deep inside the weak disorder region—and even the L^2-region—, the rate of convergence of the martingale W_n^{tr} to its limit is much different in the 2 cases: In Theorem 3.9, we have seen that the rate is polynomial in n and that the limit is

a product of W_∞ by a Gaussian r.v. In contrast, it is well known from branching theory that in the tree case,

$$\mu^{-n/2}(W_n^{\mathrm{tr}} - W_\infty^{\mathrm{tr}}) \xrightarrow{\mathrm{law}} (W_\infty^{\mathrm{tr}})^{1/2}G,$$

where $\mu > 1$ and G is a Gaussian, independent from W_∞ (for Galton-Watson process, μ is the mean offspring): the rate of convergence is exponential in this case. The convergence is much slower in the lattice case due to strong correlation between partition functions $W_n \circ \theta_{m,x}$ and $W_n \circ \theta_{m,y}$ starting from different points with same parity.

We end the chapter by mentioning some related models.

- Polymers on disordered trees: the authors in [3] study minimal subtrees supporting the free energy, and the near-critical scaling window in [181].
- Polymers on hierarchical lattices: this exactly solvable model was introduced in [81, 93]. The phase diagram was studied in [159], the intermediate disorder regime in [7], and nested critical points in [2].

Chapter 5
Semimartingale Approach and Localization Transition

The next step in our martingale analysis is to consider $\ln W_n$ as a semimartingale and to write its Doob's decomposition. Viewed as a "conditional second moment" method, this new approach is the natural continuation of the techniques from Chap. 3. However this technique was introduced much later. One concrete output is to point out polymer localization, and to relate this phenomenon to strong disorder.

5.1 Semimartingale Decomposition

It is convenient to introduce a new notation. For a sequence $(a_n)_{n\geq 0}$ (random or non-random), we set $\Delta a_n = a_n - a_{n-1}$ for $n \geq 1$. Recall (e.g., [185, 239])

Doob's Decomposition: any $(\mathscr{G}_n)$-adapted process $X = \{X_n\}_{n\geq 0} \subset L^1(\mathbb{P})$ can be decomposed in a unique way as

$$X_n = M_n(X) + A_n(X), \quad n \geq 1,$$

where $M(X)$ is an $(\mathscr{G}_n)$-martingale and $A(X)$ is predictable—i.e., $A_n(X)$ is $\mathscr{G}_{n-1}$-measurable—with $A_0 = 0$. To determine these new processes, we compute their increments

$$\Delta A_n = \mathbb{P}[\Delta X_n | \mathscr{G}_{n-1}], \qquad \Delta M_n = \Delta X_n - \mathbb{P}[\Delta X_n | \mathscr{G}_{n-1}],$$

and then $A_n = \sum_{t=1}^n \Delta A_t, M_n = X_0 + \sum_{t=1}^n \Delta M_t$. $M_n(X)$ and $A_n(X)$ are called respectively, the martingale part and compensator of the process X. If N is a

© Springer International Publishing AG 2017
F. Comets, *Directed Polymers in Random Environments*, Lecture Notes
in Mathematics 2175, DOI 10.1007/978-3-319-50487-2_5

square integrable martingale, then the compensator $A(N^2)$ of the process $N^2 = \{(N_n)^2\}_{n\geq 0} \subset L^1(\mathbb{P})$ is denoted by $\langle N \rangle_n$ and is given by

$$\Delta \langle N \rangle_n = \mathbb{P}[N_n^2 - N_{n-1}^2 | \mathscr{G}_{n-1}]$$
$$= \mathbb{P}[(\Delta N_n)^2 | \mathscr{G}_{n-1}] \,, \tag{5.1}$$

where the first line is the definition, and the second one is by orthogonality of increments. Here, we are interested in the Doob's decomposition of $X_n = -\ln W_n$, whose martingale part and the compensator will be denoted M_n and A_n respectively

$$-\ln W_n = M_n + A_n. \tag{5.2}$$

Since W_n is a martingale, $-\ln W_n$ is a submartingale, and then A_n is an increasing process: $n \mapsto A_n$ is non-decreasing for almost every ω. To compute M_n and A_n, we introduce

$$U_n = P_{n-1}^{\beta,\omega}[e^{\beta \omega(n,S_n) - \lambda(\beta)}] - 1 \,.$$

It is then clear that

$$W_n / W_{n-1} = 1 + U_n, \tag{5.3}$$

and will write W_n in the form of a telescopic product $W_n = \prod_{t=1}^n (1 + U_t)$. Hence,

$$\Delta A_n = -\mathbb{P}\big[\ln(1 + U_n)|\mathscr{G}_{n-1}\big] \,, \tag{5.4}$$
$$\Delta M_n = -\ln(1 + U_n) + \mathbb{P}\big[\ln(1 + U_n)|\mathscr{G}_{n-1}\big] \,. \tag{5.5}$$

A key role in the asymptotics of the model is played by the following random variables on $(\Omega_\omega, \mathscr{G}, \mathbb{P})$,

$$I_n = \sum_{x \in \mathbb{Z}^d} P_{n-1}^{\beta,\omega}\{S_n = x\}^2 \,. \tag{5.6}$$

We now provide an interpretation of I_n in terms of so-called replica. On the product space Ω_{path}^2, we consider the probability measure

$$[P_n^{\beta,\omega}]^{\otimes 2} = P_n^{\beta,\omega} \otimes P_n^{\beta,\omega} \,,$$

that we will view as the distribution of the couple $(S, \widetilde{S})$ with $\widetilde{S} = (\widetilde{S}_k)_{k \geq 0}$ an independent copy of $S = (S_k)_{k \geq 0}$ with law $P_n^{\beta,\omega}$. The paths S and $\widetilde{S}$ are called **replica**, they are independent polymers sharing the same environment. We can write (5.6) in the form

$$I_n = [P_{n-1}^{\beta,\omega}]^{\otimes 2}(S_n = \widetilde{S}_n) \, . \tag{5.7}$$

Hence, the summation

$$\sum_{1 \le k \le n} I_k \tag{5.8}$$

is the expected amount of overlap up to time n of two independent polymers in the same (fixed) environment. This can be viewed as an analogue to the so-called *replica overlap*, a central quantity in the context of disordered systems, e.g. mean field spin glass, and also of directed polymers on trees [94].

The large time behavior of (5.8) and the normalized partition function W_n are related as follows.

Theorem 5.1 *Let $\beta \ne 0$. Then,*

$$\{W_\infty = 0\} = \left\{ \sum_{n \ge 1} I_n = \infty \right\}, \quad \mathbb{P}\text{-}a.s. \tag{5.9}$$

Moreover, if $\mathbb{P}\{W_\infty = 0\} = 1$, there exist $c_1, c_2 \in (0, \infty)$ depending on $\beta, \mathbb{P}$, such that $\mathbb{P}$-a.s.,

$$c_1 \sum_{1 \le k \le n} I_k \le - \ln W_n \le c_2 \sum_{1 \le k \le n} I_k \quad \text{for large enough } n\text{'s,} \tag{5.10}$$

and also

$$\lim_{n \to \infty} \frac{- \ln W_n}{A_n} = 1 \quad \text{a.s.}$$

Proof (Proof of Theorem 5.1) To conclude (5.9) and (5.10), it is enough to show the following (5.11) and (5.12):

$$\{W_\infty = 0\} \subset \left\{ \sum_{n \ge 1} I_n = \infty \right\}, \quad \mathbb{P}\text{-}a.s. \tag{5.11}$$

There are $c_1, c_2 \in (0, \infty)$ such that

$$\left\{ \sum_{n \ge 1} I_n = \infty \right\} \subset \{(5.10) \text{ holds}\}, \quad \mathbb{P}\text{-}a.s. \tag{5.12}$$

In view of (5.5), and since the variance is bounded by the second moment (conditional on $\mathscr{G}_{n-1}$),

$$\mathbb{P}[(\Delta M_n)^2 | \mathscr{G}_{n-1}] \le \mathbb{P}[\ln^2(1 + U_n) | \mathscr{G}_{n-1}].$$

By (5.1), this means that

$$\Delta \langle M \rangle_n \le \mathbb{P}[\ln^2(1 + U_n) | \mathscr{G}_{n-1}]. \tag{5.13}$$

We now claim that there is a constant $c \in (0, \infty)$ such that

$$\frac{1}{c} I_n \le \Delta A_n \le c I_n \quad , \quad \Delta \langle M \rangle_n \le c I_n. \tag{5.14}$$

Indeed, both follow from (5.4), (5.13) and Lemma 5.1 below; $\{e_i\}$, $\{\alpha_i\}$ and $\mathbb{P}$ in the lemma play the roles of $\{e^{\beta\omega(n,z)-\lambda(\beta)}\}_{|z|_1 \le n}$, $\{P_{n-1}^{\beta,\omega}(S_n = z)\}_{|z|_1 \le n}$ and $\mathbb{P}[\cdot | \mathscr{G}_{n-1}]$.

We now conclude (5.11) from (5.14) as follows (the equalities and the inclusions here being understood as $\mathbb{P}$-a.s.):

$$\left\{ \sum_{n \ge 1} I_n < \infty \right\} \subset \{A_\infty < \infty, \ \langle M \rangle_\infty < \infty\}$$

$$\subset \{A_\infty < \infty, \ \lim_{n \to \infty} M_n \text{ exists and is finite}\}$$

$$\subset \{ \lim_{n \to \infty} \ln W_n \text{ exists and is finite}\}$$

$$= \{W_\infty > 0\} .$$

Here, on the second line, we have used a well-known property for martingales, e.g. [104, p. 255, (4.9)]: a square integrable martingale converges a.s. on the event $\{\langle M \rangle_\infty < \infty\}$.

Finally we prove (5.12). By (5.14), it is enough to show that

$$\{A_\infty = \infty\} \subset \left\{ \lim_{n \to \infty} -\frac{\ln W_n}{A_n} = 1 \right\}, \quad \mathbb{P}\text{-a.s.} \tag{5.15}$$

Thus, let us suppose that $A_\infty = \infty$, and consider two cases. If $\langle M \rangle_\infty < \infty$, then again by [104, p. 255, (4.9)], $\lim_{n \to \infty} M_n$ exists and is finite and therefore (5.15) holds. If, on the contrary, $\langle M \rangle_\infty = \infty$, then we will use the law of large numbers for martingales, see [104, p. 255, (4.10)]: $M_n / \langle M \rangle_n \to 0$ a.s. on the event

$\{\langle M \rangle_\infty = \infty\}$. In this case we see that

$$-\frac{\ln W_n}{A_n} = \frac{M_n}{\langle M \rangle_n} \frac{\langle M \rangle_n}{A_n} + 1 \longrightarrow 1 \quad \mathbb{P}\text{-a.s.} \tag{5.16}$$

by (5.14). This completes the proof of Theorem 5.1. $\qquad\square$

We will complete the proof by stating and proving Lemma 5.1 below. However, before doing that, we start by giving a short and direct proof of (5.14), under the additional assumption of a bounded environment. Then, the argument will be more transparent.

Proof (Proof of Lemma 5.1, When $|\omega(t,x)| \leq K$ a.s.) Under this assumption, for fixed β, U_n stays in a fixed interval $\mathcal{I}$ which is bounded away from -1 and $+\infty$, and there exist constants $C_\pm \in (0, \infty)$ such that

$$u - C_- u^2 \leq \ln(1 + u) \leq u - C_+ u^2, \quad u \in \mathcal{I}.$$

Recalling (5.4), we have in one direction:

$$\begin{aligned}
\Delta A_n &= -\mathbb{P}\big[\ln(1 + U_n)\big|\mathcal{G}_{n-1}\big], \\
&\leq -\mathbb{P}\big[U_n\big|\mathcal{G}_{n-1}\big] + C_-\mathbb{P}\big[U_n^2\big|\mathcal{G}_{n-1}\big], \\
&= C_-[P_{n-1}^{\beta,\omega}]^{\otimes 2}\mathbb{P}\Big[(e^{\beta\omega(n,S_n)-\lambda(\beta)} - 1)(e^{\beta\omega(n,\widetilde{S}_n)-\lambda(\beta)} - 1)\Big] \\
&= C_-\sum_x [P_{n-1}^{\beta,\omega}]^{\otimes 2}(S_n = \widetilde{S}_n = x)\mathbb{P}[(e^{\beta\omega(n,x)-\lambda(\beta)} - 1)^2] \\
&= C_-(e^{\lambda_2(\beta)} - 1)\sum_x [P_{n-1}^{\beta,\omega}]^{\otimes 2}(S_n = \widetilde{S}_n = x) \\
&= \mathrm{Cst}\, I_n,
\end{aligned}$$

with λ_2 defined in (3.12). Similarly, one gets the other direction $\Delta A_n \geq \mathrm{Cst}\, I_n$, which proves the first two inequalities claimed in (5.14). It is clear that $\ln^2(1 + u) \leq Cu^2$ for $u \in \mathcal{I}$ with some finite constant C, which is enough to get the last claim in (5.14). $\qquad\square$

We now gives the lemma, which takes care of the general case.

Lemma 5.1 *Let e_i, $1 \leq i \leq m$ be positive, non-constant i.i.d. random variables on a probability space $(H, \mathcal{G}, \mathbb{P})$ such that*

$$\mathbb{P}[e_1] = 1, \quad \mathbb{P}[e_1^3 + \ln^2 e_1] < \infty.$$

For $\{\alpha_i\}_{1 \leq i \leq m} \subset [0, \infty)$ such that $\sum_{1 \leq i \leq m} \alpha_i = 1$, define a centered random variable $U > -1$ by $U = \sum_{1 \leq i \leq m} \alpha_i e_i - 1$. Then, there exists a constant $c \in (0, \infty)$, independent of m and of $\{\alpha_i\}_{1 \leq i \leq m}$, such that

$$\frac{1}{c} \sum_{1 \leq i \leq m} \alpha_i^2 \leq \mathbb{P}\left[\frac{U^2}{2 + U}\right], \tag{5.17}$$

$$\frac{1}{c} \sum_{1 \leq i \leq m} \alpha_i^2 \leq -\mathbb{P}\left[\ln(1 + U)\right] \leq c \sum_{1 \leq i \leq m} \alpha_i^2, \tag{5.18}$$

$$\mathbb{P}\left[\ln^2(1 + U)\right] \leq c \sum_{1 \leq i \leq m} \alpha_i^2. \tag{5.19}$$

Proof Proof in the general case, i.e., under assumption (1.1): In this proof, we let $c_1, c_2, \dots$ stand for constants which are independent of $\{\alpha_i\}_{1 \leq i \leq m}$. We have by direct computations that

$$\mathbb{P}[U^2] = c_1 \sum_{1 \leq i \leq m} \alpha_i^2, \quad \mathbb{P}[U^3] \leq c_2 \sum_{1 \leq i \leq m} \alpha_i^2.$$

Then, (5.17) is obtained as follows;

$$c_1 \sum_{1 \leq i \leq m} \alpha_i^2 = \mathbb{P}\left[\frac{U}{\sqrt{2 + U}} U\sqrt{2 + U}\right]$$

$$\leq \mathbb{P}\left[\frac{U^2}{2 + U}\right]^{1/2} \mathbb{P}\left[2U^2 + U^3\right]^{1/2}$$

$$\leq c_3 \mathbb{P}\left[\frac{U^2}{2 + U}\right]^{1/2} \left(\sum_{1 \leq i \leq m} \alpha_i^2\right)^{1/2}.$$

To prove the other inequalities, it is convenient to define a function $\varphi : (-1, \infty) \to [0, \infty)$ by $\varphi(u) = u - \ln(1 + u)$, so that

$$-\mathbb{P}\left[\ln(1 + U)\right] = \mathbb{P}\left[\varphi(U)\right].$$

Since $\frac{1}{4} \frac{u^2}{2+u} \leq \varphi(u)$, $u > -1$, the left-hand-side inequality of (5.18) follows from (5.17). The right-hand-side inequality can be seen as follows. We have for any $\varepsilon \in (0, 1)$,

$$\mathbb{P}\left[\varphi(U)\right] = \mathbb{P}[\varphi(U); 1 + U \geq \varepsilon] + \mathbb{P}[\varphi(U); 1 + U \leq \varepsilon]$$

$$\leq \mathbb{P}[\varphi(U); 1 + U \geq \varepsilon] - \mathbb{P}[\ln(1 + U); 1 + U \leq \varepsilon].$$

Since $\varphi(u) \leq \frac{1}{2}(u/\varepsilon)^2$ if $1 + u \geq \varepsilon$,

$$\mathbb{P}[\varphi(U); 1 + U \geq \varepsilon] \leq \frac{1}{2}\varepsilon^{-2}\mathbb{P}[U^2]$$

$$= \frac{1}{2}\varepsilon^{-2}c_1 \sum_{1 \leq i \leq m} \alpha_i^2. \tag{5.20}$$

We now set $\gamma = -\mathbb{P}[\ln e_1] \geq 0$ and choose $\varepsilon > 0$ so small that $\ln(1/\varepsilon) - \gamma \geq 1$. We introduce another centered random variable $V = \sum_{1 \leq i \leq m} \alpha_i(\ln e_i + \gamma)$. We then see from Jensen's inequality that

$$\{1 + U \leq \varepsilon\} = \{V - \gamma \leq \ln(1 + U) \leq \ln \varepsilon\}$$

$$\subset \{-\ln(1 + U) \leq -V + \gamma\} \cap \{1 \leq -V\}.$$

Hence we have

$$-\mathbb{P}[\ln(1 + U); 1 + U \leq \varepsilon] \leq \mathbb{P}[-V; 1 \leq -V] + \gamma\mathbb{P}\{1 \leq -V\}$$

$$\leq (1 + \gamma)\mathbb{P}[V^2]$$

$$= c_4 \sum_{1 \leq i \leq m} \alpha_i^2.$$

This, together with (5.20) proves the right-hand-side inequality of (5.18). The proof of (5.19) is similar. Indeed, since $|\ln(1 + u)| \leq \varepsilon^{-1}\ln(\varepsilon^{-1})|u|$ if $\varepsilon \leq 1 + u$, we have that

$$\mathbb{P}[\ln^2(1 + U); \varepsilon \leq 1 + U] \leq \varepsilon^{-2}\ln^2(\varepsilon^{-1})\mathbb{P}[U^2].$$

We see on the other hand that

$$\{1 + U \leq \varepsilon\} = \{V - \gamma \leq \ln(1 + U) \leq \ln \varepsilon\}$$

$$\subset \{\ln^2(1 + U) \leq 2V^2 + 2\gamma^2\} \cap \{1 \leq -V\}.$$

Therefore, we obtain

$$\mathbb{P}[\ln^2(1 + U); 1 + U \leq \varepsilon] \leq 2\mathbb{P}[V^2] + 2\gamma^2\mathbb{P}\{1 \leq -V\}$$

$$\leq c_5 \sum_{1 \leq i \leq m} \alpha_i^2.$$

$\square$

Corollary 5.1 $\mathbb{P}$-*a.s.*,

$$p(\beta) = \lim_{n\to\infty} \frac{1}{n} \sum_{t=1}^{n} \mathbb{P}\left[\ln P_{t-1}^{\beta,\omega}[e^{\beta\omega(t,S_t)}]\Big|\mathcal{G}_{t-1}\right]$$

Proof This follows directly from (5.16).   □

Observe also that $p_n(\omega;\beta) = (1/n)\sum_{t=1}^{n} \ln(Z_t/Z_{t-1})$, and recall that $p_n(\omega;\beta)$ converge in L^1 to a deterministic limit. We have

$$p(\beta) = \lim_{n\to\infty} \frac{1}{n} \sum_{t=1}^{n} \mathbb{P}\left[\ln P_{t-1}^{\beta,\omega}[e^{\beta\omega(t,S_t)}]\right] = \text{a.s.}-\lim_{n\to\infty} \frac{1}{n} \sum_{t=1}^{n} \ln P_{t-1}^{\beta,\omega}[e^{\beta\omega(t,S_t)}].$$

5.2   Weak Disorder and Diffusive Regime

Assume weak disorder holds. Taking the limit $n \to \infty$ in (2.4), we see that the sequence of Markov chains defined by the polymer measure of length n has an infinite volume limit: For almost every ω, it converges as $n \to \infty$ to a (time-inhomogeneous) Markov chain on $\mathbb{Z}^d$ with transitions

$$P_{\infty}^{\beta,\omega}(S_{i+1} = y|S_i = x) = e^{\beta\omega(i+1,y)-\lambda(\beta)} \frac{W_{\infty} \circ \theta_{i+1,y}}{W_{\infty} \circ \theta_{i,x}} P(S_{i+1} = y|S_i = x).$$

In the L^2-region, Moreno Flores [175] studies the environment process $\widehat{\omega}_n$,

$$\widehat{\omega}_n(t,x) = \omega(n+t, S_n + x)$$

seen by the moving particle. Let $\mathbb{P}_n$ be the law of $\widehat{\omega}_n$ under $\mathbb{P} \times P_n^{\beta,\omega}$—so-called, law of the environment seen from the particle. Then, the density of $\mathbb{P}_n$ relative to $\mathbb{P}$ is given by

$$\frac{d\mathbb{P}_n}{d\mathbb{P}} = \sum_{y\in\mathbb{Z}^d} P_n^{\beta,\theta_{-n,-y}\omega}(S_n = y),$$

see [175, Eqs. (8)–(10)]. Using the local limit Theorem 3.5, it is shown in this paper that, in the L^2-region the density converges

$$\frac{d\mathbb{P}_n}{d\mathbb{P}} \longrightarrow e^{\beta\omega(0,0)-\lambda(\beta)} \times W_{\infty} \circ \overleftarrow{\theta}_{0,0}$$

in $L^1(\mathbb{P})$ as $n \to \infty$.

Remark 5.1 (Local Limit Theorem and Overlap) It is quite instructive to see that, at least at a heuristic level, the local limit theorem for the polymer measure (see the statement in Theorem 3.5) implies weak disorder. Indeed, we have

$$I_n = \sum_{x \in \mathbb{Z}^d} P_n^{\beta,\omega}(S_n = x)^2 \qquad \text{(by definition)}$$

$$\simeq \sum_{x \in \mathbb{Z}^d} \left[W_\infty \circ \theta_{n,x}^{\leftarrow} \times P(S_n = x) \right]^2 \qquad \text{(local limit theorem)}$$

$$\simeq \mathbb{P}(W_\infty^2) \times \sum_{x \in \mathbb{Z}^d} P(S_n = x)^2 \qquad \text{(ergodic theorem)}$$

$$\simeq Cn^{-d/2},$$

which is the general term of a summable series for $d \geq 3$. The local limit theorem has been proved in Theorem 3.5 under the L^2 condition (3.12). One could also justify the third line by a second moment computation for small β. The above calculation indicates that the validity of local limit theorem is a natural definition for the polymer to be in the weak disorder regime (better than the central limit theorem itself).

At a rigorous level, only a slower polynomial decay has been so far achieved for I_n [(1.17) in [78]].

Observation: (weak disorder region). In the weak disorder region, it is not difficult to see that the polymer measure is very similar to the simple random walk. Indeed, when $W_\infty > 0$, for any $A_n \in \mathscr{F}_n$ such that $P(A_n) \to 1$ as $n \to \infty$, we have

$$P_n^{\beta,\omega}(A_n) \longrightarrow 1 \quad \text{in } \mathbb{P} - \text{probability}.$$

This follows from

$$P_n^{\beta,\omega}(A_n^c) = W_n^{-1} P(e^{\beta H_n - n\lambda(\beta)}; A_n^c) \longrightarrow W_\infty^{-1} \times 0 = 0$$

in $\mathbb{P}$-probability, since $P(e^{\beta H_n - n\lambda(\beta)}; A_n^c) \to 0$ in L^1-norm.

This applies for instance to the set $A_n = \{|S_n| \in [a_n, b_n]\}$ with any positive sequences a_n, b_n such that $a_n = o(n^{1/2}), n^{1/2} = o(b_n)$. This shows that the polymer does not spread out much more than the simple random walk.

It is natural to expect that diffusive behavior takes place not only in the perturbative regime—i.e., under the stronger assumption (3.12)—but in the full weak disorder region. This would be the final statement, and fortunately, it can be proved here:

Theorem 5.2 ([75]) *Assume $d \geq 3$ and weak disorder (3.5). Then, for all bounded continuous function F on the path space,*

$$\lim_n P_n^{\beta,\omega}[F(S^{(n)})] = \mathbf{E}F(B)$$

in probability, where $S^{(n)}$ is the rescaled path defined by $S^{(n)} = (S_{nt}/\sqrt{n})_{t \geq 0}$ and B is the Brownian motion with diffusion matrix $d^{-1}I_d$. In particular, this holds for all $\beta \in [0, \bar{\beta}_c)$.

In the proof of Theorem 5.2 convergence of the series $\sum I_n$ is used as a main technical quantitative ingredient. The reader is referred to [75] for the (rather intricate) proof.

Exponents: Incidentally, we see that the scaling relation between exponents does hold in the full weak disorder region, with $\chi^{\perp} = 1/2$ and $\chi^{\parallel} = 0$.

5.3   Bounds on the Critical Temperature by Size-Biasing

Recall definitions (3.13)–(4.8). So far we have obtained the bounds

$$\beta_{L^2} \leq \bar{\beta}_c \leq \beta_c \leq \beta_c^{tr}$$

for the critical temperature inverse. Our aim here is to improve the knowledge of $\bar{\beta}_c$. It has been conjectured [174] that $\beta_c = \beta_{L^2}$, but we will see it is wrong. The second moment of the partition function gave us the lower bound, and higher moments, including integer ones which are explicit, will not help us. To improve the lower bound, the way forward is to estimate fractional moments (but this time, $\mathbb{P}[W_n^\alpha]$ for $\alpha \in (1, 2)$). Using convexity arguments, sufficient conditions (based on entropy controls) ensuring bounded fractional moments were introduced in [107] and systematically studied in [55].

In this section we describe a smart approach introduced by Birkner [38] based on size-biasing. Define

$$\beta_{sb} = \sup \left\{ \beta \geq 0 : P^{\otimes 2} \left[\exp\{\lambda_2(\beta)N_\infty(S, \widetilde{S})\} | \widetilde{S} \right] < \infty \quad \widetilde{S} - a.s. \right\}. \tag{5.21}$$

Observe that the event $\{P^{\otimes 2}[\exp\{\beta N_\infty(S, \widetilde{S})\} | \widetilde{S}] < \infty\}$ belongs to the tail sigma-field of $\widetilde{S}$, and therefore it has probability 0 or 1 according to Kolmogorov. Thus β_{sb} is well defined in $[0, \infty]$, and one easily sees that $\beta_{sb} \geq \beta_{L^2}$.

Theorem 5.3 ([38]) *Weak disorder holds when* $\beta < \beta_{sb}$. *Thus,*

$$\beta_{sb} \leq \bar{\beta}_c \leq \beta_c.$$

Before starting the proof, we introduce the size-bias partition function, and first the necessary notations. We will deal with two copies $S, \widetilde{S}$ of the path and two instances $\omega, \widehat{\omega}$ of the environment. All four will be independent, and for consistency, we will denote by $P, \widetilde{P}, \mathbb{P}, \widehat{\mathbb{P}}$ the corresponding (marginal) probabilities and expectations respectively. E.g., we will use, in this section only, the notation in the right-hand side for

$$P^{\otimes 2}\big[\exp\{\lambda_2(\beta)N_\infty(S,\widetilde{S})\}\big|\widetilde{S}\big] = P\big[\exp\{\lambda_2(\beta)N_\infty(S,\widetilde{S})\}\big],$$

which is a $\widetilde{S}$-measurable random variable. We also use the short notations

$$e = e(\omega) = \big(e(i,x); i \in \mathbb{N}, x \in \mathbb{Z}^d\big), \quad e(i,x) = e(i,x;\omega) = \exp\{\beta\omega(i,x) - \lambda(\beta)\}.$$

Let $\widehat{\omega}$ be an i.i.d. environment, independent of ω, with the size-biased distribution

$$\widehat{\mathbb{P}}\big(\widehat{\omega}(i,x) \in \cdot\big) = \mathbb{P}\big[e(i,x); \omega(i,x) \in \cdot\big].$$

The corresponding weight is $\widehat{e}(i,x) = \exp\{\beta\widehat{\omega}(i,x) - \lambda(\beta)\}$. Given a path $\widetilde{S}$ in addition to $\omega, \widehat{\omega}$, we define a new family of weights $\widehat{e}_{\widetilde{S}} = \big(\widehat{e}_{\widetilde{S}}(i,x); i \in \mathbb{N}, x \in \mathbb{Z}^d\big)$ with

$$\widehat{e}_{\widetilde{S}}(i,x) = \begin{cases} \widehat{e}(i,x) \text{ if } \widetilde{S}_i = x, \\ e(i,x) \text{ if } \widetilde{S}_i \neq x. \end{cases} \tag{5.22}$$

Then, the variable

$$\widehat{W}_n^{e,\widehat{e},\widetilde{S}} = P\left[\prod_{i=1}^n \widehat{e}_{\widetilde{S}}(i,S_i)\right],$$

is called the size-biased version of $W_n \equiv P\big[\prod_i e(i,S_i)\big]$ for the following reason:

Proposition 5.1 *For $f : [0,\infty) \to \mathbb{R}$ bounded measurable,*

$$\mathbb{P}\big[W_n f(W_n)\big] = \mathbb{P}\widehat{\mathbb{P}}\widetilde{P}\big[f(\widehat{W}_n^{e,\widehat{e},\widetilde{S}})\big].$$

Proof To prove the proposition we compute

$$\mathbb{P}\big[W_n f(W_n)\big] = \mathbb{P}\left[\widetilde{P}\big[\prod_i e(i,\widetilde{S}_i)\big] f(P\big[\prod_i e(i,S_i)\big])\right]$$

$$= \widetilde{P}\left[\mathbb{P}\left[\big[\prod_i e(i,\widetilde{S}_i)\big] f(P\big[\prod_i e(i,S_i)\big])\right]\right]$$

$$= \widetilde{P}\left[(\mathbb{P}\widehat{\mathbb{P}})\left[f(P\big[\prod_i \widehat{e}_{\widetilde{S}}(i,S_i)\big])\right]\right]$$

$$\stackrel{(5.22)}{=} \mathbb{P}\widehat{\mathbb{P}}\widetilde{P}\left[f(\widehat{W}_n^{e,e,\widehat{\widetilde{S}}})\right],$$

yielding the desired result. $\qquad\qquad\qquad\qquad\qquad\qquad\qquad\qquad\qquad\qquad$ □

We now prove Theorem 5.3.

Proof It follows from the proposition that

$$(W_n)_n \text{ is uniformly integrable} \iff (\widehat{W}_n^{e,e,\widehat{\widetilde{S}}})_n \text{ is tight.}$$

On the other hand, it is well known, e.g. [75, Proposition 3.1] that

$$(W_n)_n \text{ is uniformly integrable} \iff W_\infty > 0 \text{ a.s.}$$

Since

$$(\mathbb{P}\widehat{\mathbb{P}})\big[\widehat{W}_n^{e,e,\widehat{\widetilde{S}}}\big] = P\big[\exp\{\lambda_2(\beta)N_n(S,\widetilde{S})\}\big],$$

assuming $P\big[\exp\{\lambda_2(\beta)N_\infty(S,\widetilde{S})\}\big] < \infty$ for $\widetilde{P}$-a.e. $\widetilde{S}$ is enough to imply tightness of the sequence $\widehat{W}_n^{e,e,\widehat{\widetilde{S}}}$. $\qquad\qquad\qquad\qquad\qquad\qquad\qquad$ □

We now give a corollary. Claims (i) follow from the dichotomy in Theorem 3.1, and the existence of the limit and the equivalence in (ii) are proved in [137] for Gaussian environment.

Corollary 5.2 *(i)* *Characterization of weak and strong disorder:*

$$\widehat{W}_n^{e,e,\widehat{\widetilde{S}}} \xrightarrow{\text{prob.}} +\infty \iff \text{strong disorder}$$

$$\widehat{W}_n^{e,e,\widehat{\widetilde{S}}} \text{ converges in law} \iff \text{weak disorder .}$$

(ii) Characterization of low and high temperature (for Gaussian ω):

$$\lim_{n\to\infty} \frac{1}{n}\mathbb{P}\widehat{\mathbb{P}}\widetilde{P}\left[\ln \widehat{W}_n^{e,e,\widehat{\widetilde{S}}}\right] > 0 \iff p(\beta) < \lambda(\beta)$$

$$\lim_{n\to\infty} \frac{1}{n}\mathbb{P}\widehat{\mathbb{P}}\widetilde{P}\left[\ln \widehat{W}_n^{e,e,\widehat{\widetilde{S}}}\right] = 0 \iff p(\beta) = \lambda(\beta).$$

Example 5.1 (Gaussian Environment) Assume $\omega(t,x) \sim \mathcal{N}(0,1)$. Then, the size-biased environment $\widehat{\omega}$ is $\mathcal{N}(\beta,1)$-distributed. Writing such variables $\widehat{\omega}(i,x) = \beta + \omega'(i,x)$ with i.i.d. $\omega'(t,x) \sim \mathcal{N}(0,1)$ defined by this equality, we can express

$$\widehat{W}_n^{e,e,\widehat{\widetilde{S}}} = P\left[\left(\prod_{i=1}^{n} e'_{\widetilde{S}}(i,S_i)\right) \times \exp\{\lambda_2(\beta)N_n(S,\widetilde{S})\}\right]$$

in term of the weights e' associated to ω' via the construction (5.22). In the right-hand side the weights now are all $\mathcal{N}(0,1)$-distributed. In this form the model looks like a polymer with disorder on sites and pinning with constant reward on a random path.

Remark 5.2

(i) Relation with the pinning model: As pointed out in [39], there is an interesting connection between the gap $\beta_{L^2} < \beta_{sb}$ and the question of disorder relevance versus irrelevance in pinning model with disorder. The reader is referred to Sect. 12.7.3. in [92].

(ii) As mentioned above, it has been conjectured [174] in the physics literature that $\beta_c = \beta_{L^2}$ in dimension 3 and larger. This was first disproved in [55] for special environments. Then, it was disproved for general environment in dimension $d \geq 5$ [42], later in dimension 4 [39] and for $d = 3$ in [30] in the discrete case and [40] in the continuous case. Section 1.4 in [39] provides a detailed discussion of the relations with disordered pinning on a random walk, and also points out that earlier results on the critical point shift for the pinning model [96] was contradicting the conjecture.

5.4 Localization Versus Delocalization

In the strong disorder phase the attraction to regions in the environment with low energy has a non-negligible contradicting effect on the entropic tendency of the polymer to diffuse, resulting in a localization (pinning effect) of the polymer to a region in the environment where the balance between entropy and energy is optimal. In this phase, the transversal fluctuations of the polymer are mainly due to the fluctuations in the shape of this optimal region. This pinning becomes absolute in the extreme case of zero temperature (formally, the weak limit of the Gibbs measure for infinite β). At zero temperature, entropy no longer plays a part and the system is in one of its ground states—i.e., states in which the polymer path is a minimizer of the energy.

We want to characterize the following phenomenon which can be observed experimentally and numerically: For large β the polymer concentrates around the n-geodesics, i.e. the maximizers of H_n. For instance we could try to study $P_n^{\beta,\omega}(S \in G_n)$ for G_n a neighborhood of the set of the n-geodesics. A difficulty is that little is known on the geodesics. For the case $d = 1$ the reader can refer to Part 1 of [186]. We will also reduce our ambition in considering only the ending point of the path. A simpler quantity is the random variable J_n, which is the **probability of the favourite endpoint** for the polymer of size n,

$$J_n = \max_{x \in \mathbb{Z}^d} P_{n-1}^{\beta,\omega}\{S_n = x\} . \tag{5.23}$$

Indeed, J_n is small when the measure is spread out—for instance if $\beta = 0$, $J_n = \mathcal{O}(n^{-d/2})$—but J_n should be much larger when $P_n^{\beta,\omega}$ concentrates on a small number of paths ($J_n \in (0,1]$). The advantage is that we don't need to know where favourite point(s) is (are) located ! The shift in the time index is harmless up to a constant factor, we could have taken in the definition on I_n the maximum of $P_{n-1}^{\beta,\omega}\{S_{n-1} = x\}$ without changing its essence, but the present one is more natural.

In fact, J_n can be compared to $I_n = \sum_x P_{n-1}^{\beta,\omega}(S_n = x)^2$ from (5.6),

$$J_n^2 \le I_n \le J_n , \tag{5.24}$$

as can be seen by keeping only the biggest term in the sum for the lower bound, and using that $\sum_x P_{n-1}^{\beta,\omega}(S_n = x) = 1$ for the upper bound. It follows that, I_n vanishes if and only if J_n does.

In view of the above discussion, the following definition from [58, 78], is most natural:

Definition 5.1 We say that the polymer is **localized** if

$$\liminf_{n\to\infty} \frac{1}{n} \sum_{t=1}^{n} J_t > 0, \quad \mathbb{P}\text{-a.s.} \tag{5.25}$$

and that the polymer is **delocalized** if

$$\lim_{n\to\infty} \frac{1}{n} \sum_{t=1}^{n} J_t = 0 \quad \mathbb{P}\text{-a.s.} \tag{5.26}$$

Roughly, delocalization and localization correspond to J_n vanishing or not as $n \to \infty$. The following result shows that necessarily one of the two cases happens. But it is all the more a *criterion* for localization and delocalization.

Theorem 5.4 (Localization Transition) *Let $\beta \neq 0$. The polymer is*

- *localized if and only if $p < \lambda$,*
- *delocalized if and only if $p = \lambda$.*

Localization occurs for all $\beta > \beta_c$, and delocalization for $\beta \leq \beta_c$.

Proof In view of (5.24), Theorem 5.4 directly follows from Theorem 5.1, (5.10), from the inequality

$$\left(\frac{1}{n} \sum_{t=1}^{n} J_t \right)^2 \leq \frac{1}{n} \sum_{t=1}^{n} J_t^2$$

and from the definition of β_c. $\square$

Remark 5.3 From this and from (5.24), we observe that I_n and J_n have Cesaro limit of the same nature. Either $n^{-1} \sum_{t=1}^{n} I_t$ and $n^{-1} \sum_{t=1}^{n} J_t$ have both a.s. positive limits (superior and inferior), or they vanishes a.s. as $n \to \infty$.

At this point we recall well known facts [219] for the simple random walk, i.e. the behavior of I_n and J_n in the case $\beta = 0$:

$$\max_{x \in \mathbb{Z}^d} P\{S_n = x\} = \mathcal{O}(n^{-d/2}) , \tag{5.27}$$

$$P^{\otimes 2}\{S_n = \widetilde{S}_n\} = \mathcal{O}(n^{-d/2}) , \tag{5.28}$$

as $n \to \infty$. The decay rate $n^{-d/2}$ in (5.27) can be understood as the position of S_n being roughly uniformly distributed over the Euclidean ball in $\mathbb{Z}^d$ with radius const. $\times \sqrt{n}$.

For $\beta \neq 0$ but in the weak disorder region, we first note from the convergence of $\sum I_n$ that $J_n \to 0$. In this region, can still prove (5.27) in some specific models, only in a weaker form with a smaller exponent—see e.g. (1.17) in [78]. Anyway the picture remains similar, with the position S_n of the polymer being widely spread out, or "delocalized".

Remark 5.4 For the Parabolic Anderson model (PAM), [60] proves that localization occurs at strong disorder, everywhere in the sense:

$$W_\infty = 0 \implies \limsup_t \max_x \mathbb{P}_t^{\beta,\omega}(S_t = x) > c$$

for some positive constant c. This completes the statement that $W_\infty > 0$ is equivalent to diffusivity.

Remark 5.5 (Atoms of Endpoints at Low Temperature) In a recent preprint [25] Bates and Chatterjee prove the characterization of low temperature by the property that the endpoint distribution is asymptotically purely atomic, as defined in [235].

β	0 L^2 region	β_{L^2}	$\bar{\beta}_c$	β_c	∞
regime	weak disorder, diffusive			strong disorder	
polymer behavior	delocalized			localized	
I_n	$\sum I_n < \infty$			$\sum I_n = \infty$	
J_n	$\simeq n^{-a}$	$o(1)$	non summable	J_n is $\mathcal{O}(1)$	
Parabolic Anderson	$\lim J_t = 0$			$\limsup J_t > c$ a.s.	

Fig. 5.1 Different regimes for the polymer ($d \geq 3$; $a \in (0, d/2]$). For different ranges of values for β, we summarize the regime and the behavior for the polymer, and we indicate the order of magnitude of I_n and J_n

By definition, this property means that, for all sequence $\varepsilon_i \to 0^+$,

$$\lim_{n \to \infty} n^{-1} \sum_{i=1}^{n} P_i^{\beta,\omega}(S_i \in B_i^{\varepsilon_i}) = 1 \tag{5.29}$$

in probability, where

$$B_i^{\varepsilon} = \left\{ x \in \mathbb{Z}^d : P_i^{\beta,\omega}(S_i = x) > \varepsilon \right\}$$

is the set of all atoms larger than $\varepsilon > 0$. The quantity $P_i^{\beta,\omega}(S_i \in B_i^{\varepsilon})$ is the mass for the polymer of size i of the ε-atoms, i.e., of the sites with mass larger than ε. It is close to one if the total mass of the endpoint under the polymer measure is mostly concentrated on sites of mass larger than ε, and it is close to 0 if the mass is scattered over many points of small mass. A deep result of [25] is that

- $\beta > \beta_c$ implies (5.29),
- $\beta \leq \beta_c$ implies $\lim_{n\to\infty} n^{-1} \sum_{i=1}^{i} P_i^{\beta,\omega}(S_i \in B_i^{\varepsilon_i}) = 0$ for a sequence $\varepsilon_i \to 0$.

In Theorem 7.4 we will discuss (with complete proofs) a specific example where localization occurs on a single blob, i.e., all the atoms are grouped.

We summarize the whole discussion in the table of Fig. 5.1.

Chapter 6
The Localized Phase

In this chapter we start to analyse the behavior of the polymer in its localized regime. We will make precise the image of the corridors where the polymer wants to be. We will see that localization happens at all temperature in dimension $d = 1$ and 2. We illustrate the phenomenon by simulation experiments. Finally, a precise picture will be achieved when the environment has heavy tails. However, we leave some matter on the localized phase for the forthcoming Sects. 7.4 and 9.7.

6.1 Path Localization

The next step in understanding localization is to have results for the whole polymer path, not only for the endpoint. The goal is to describe the corridors where the polymer is pinned. In the localized phase, we can construct a "favourite path" (dependent on the realization of the medium and the temperature) such that, asymptotically, the proportion of time the polymer spends together with this path is strictly positive. Moreover, the limit approaches its maximum value as the temperature vanishes. See [69] for the parabolic Anderson model and [77] for Brownian motion in Poissonian medium.

In this section, we stick to discrete time, making the arguments most transparent. The main ingredient being integration by parts, we need special laws for the environment. Consider the usual model (1.2), with a Gaussian environment

$$\omega(t, x) \sim \mathcal{N}(0, 1) .$$

For $\mathbf{y} = (y_t)_t : \mathbb{N} \to \mathbb{Z}^d$ and S a path, we extend N_n from (3.14) by

$$N_n(S, \mathbf{y}) = \sum_{t=1}^{n} \mathbf{1}_{S_t = y_t} ,$$

© Springer International Publishing AG 2017
F. Comets, *Directed Polymers in Random Environments*, Lecture Notes
in Mathematics 2175, DOI 10.1007/978-3-319-50487-2_6

the number of intersections between S and $\mathbf{y}$ up to time n. Define the parameter region

$$\mathscr{I} = \{\beta > 0 : p \text{ is differentiable at } \beta, p'(\beta) < \lambda'(\beta)\}. \tag{6.1}$$

By convexity, the set where p is not differentiable is at most countable. The set $\mathscr{I}$ is a subset of the (closure of the) low temperature region. It is conjectured that these two sets coincide.

Conjecture 6.1

$$\mathscr{I} = \{\beta > 0 : p(\beta) < \lambda(\beta)\} = (\beta_c, \infty).$$

Let us state a path localization result, which did not appear so far in the literature for the discrete model. It builds over the techniques of [69, 77] in our simple setup.

Theorem 6.1 *Assume that the environment is Gaussian. There exists* $\mathbf{y}^{(n)} : [0, n] \to \mathbb{Z}^d$ *such that*

$$\liminf_{n \to \infty} \mathbb{P}P_n^{\beta,\omega} \left[\frac{N_n(S, \mathbf{y}^{(n)})}{n} \right] \geq 1 - \frac{p'}{\lambda'}(\beta) > 0, \tag{6.2}$$

for all $\beta \in \mathscr{I}$. *Moreover,*

$$\lim_{\beta \to \infty} \liminf_{n \to \infty} \mathbb{P}P_n^{\beta,\omega} \left[\frac{N_n(S, \mathbf{y}^{(n)})}{n} \right] = 1. \tag{6.3}$$

It is important to make a few comments:

- In view of its definition (6.4) we call $\mathbf{y}^{(n)}$ the *favourite path* for the polymer, even though it is not a path for the random walk—it has jumps. It depends on all the parameters β, ω, n and we don't have much information on it. However, the first claim shows that its neighborhood is a corridor where the polymer path likes to be, and is worth the name of *path localization*. The first claim shows that the polymer spends a positive proportion of the time at the same place as the favourite path. Hence it has to remain more or less close to it. The neighborhood of the favourite path $\mathbf{y}^{(n)}$ from (6.4) is the natural candidate for describing the corridor where the polymer path is pinned.
- Since N_n/n is less than 1, the second claim, so-called *complete localization*, states that this preference becomes overwhelming as β grows.

Proof

Step 1: Since the environment is Gaussian, we can use the integration by parts formula of Lemma A.3. We compute

$$\frac{d}{d\beta} \mathbb{P}p_n(\omega, \beta) = \mathbb{P}\frac{d}{d\beta}p_n(\omega, \beta)$$

$$= \frac{1}{n} \sum_{t \leq n, x} \mathbb{P}\left[P_n^{\beta, \omega}(S_t = x)\omega(t, x)\right]$$

$$\overset{(A.8)}{=} \frac{\beta}{n} \sum_{t \leq n, x} \mathbb{P}\left[P_n^{\beta, \omega}(S_t = x) - P_n^{\beta, \omega}(S_t = x)^2\right]$$

$$= \beta \left(1 - \mathbb{P}P_n^{\beta, \omega \otimes 2}\left[\frac{N_n(S, \widetilde{S})}{n}\right]\right).$$

Recall that the sequence $\mathbb{P}p_n(\omega, \beta)$ of convex functions converges to $p(\beta)$. When $\beta \in \mathscr{I}$, the limit is differentiable at β, and a standard result on convex functions implies the existence and the value of the following limit,

$$\lim_{n \to \infty} \mathbb{P}P_n^{\beta, \omega \otimes 2}\left[\frac{N_n(S, \widetilde{S})}{n}\right] = 1 - \frac{p'(\beta)}{\beta} = 1 - \frac{p'}{\lambda'}(\beta)$$

since $\lambda(\beta) = \beta^2/2$ in the Gaussian case.

Step 2: (2-to-1 step) For fixed n, β, ω, define

$$\mathbf{y}^{(n)}(t) = \arg\max_{x \in \mathbb{Z}^d} P_n^{\beta, \omega}(S_t = x), \quad t = 1, 2 \ldots, n. \tag{6.4}$$

By definition,

$$P_n^{\beta, \omega \otimes 2}(S_t = \widetilde{S}_t) = \sum_x P_n^{\beta, \omega}(S_t = x)^2 \leq P_n^{\beta, \omega}(S_t = \mathbf{y}_t^{(n)}),$$

so we obtain

$$\mathbb{P}P_n^{\beta, \omega \otimes 2}\left[\frac{N_n(S, \widetilde{S})}{n}\right] \leq \mathbb{P}P_n^{\beta, \omega}\left[\frac{N_n(S, \mathbf{y}^{(n)})}{n}\right],$$

and the claim (6.2) follows from the first step.

Step 3: Recalling the improved annealed bound (2.16), we see that the free energy grows linearly in β, which implies that $p'(\beta) \leq C < \infty$ by convexity. Thus, the right-hand side of (6.2) is larger than $1 - C/\beta$, yielding the desired claim (6.3).

$\square$

6.2 Low Dimensions

Dimensions $d = 1$ and 2 are special, due to recurrence of the simple random walk— more precisely, due to recurrence of the difference $S_n - \widetilde{S}_n$ under the product measure $P^{\otimes 2}$, which enforces the interactions between the polymer and its environment. One of our main results here is that strong disorder holds for all non-zero β in dimensions 1 and 2.

Theorem 6.2 *Assume $d = 1$ or $d = 2$. For all $\beta \neq 0$, $p(\beta) < \lambda(\beta)$ and therefore $W_\infty = 0$.*

The result $W_\infty = 0$ is due to [58, 78]; the stronger statement $p(\beta) < \lambda(\beta)$ is from [157].

6.2.1   Overlap Estimates

Here is an elementary account for $W_\infty = 0$. Overlap estimates can be derived using on the following elementary observation. For all $z \in \mathbb{Z}^d$,

$$[P_{t-1}^{\beta,\omega}]^{\otimes 2}(S_t = \widetilde{S}_t + z) = \sum_x P_{t-1}^{\beta,\omega}(S_t = x)P_{t-1}^{\beta,\omega}(S_t = x + z)$$

$$\leq \left(\sum_x P_{t-1}^{\beta,\omega}(S_t = x)^2 \times \sum_x P_{t-1}^{\beta,\omega}(S_t = x + z)^2\right)^{1/2}$$

$$= [P_{t-1}^{\beta,\omega}]^{\otimes 2}(S_t = \widetilde{S}_t)$$

$$= I_t$$

where the inequality is from Cauchy-Schwarz.

Proof Proof of the claim $W_\infty = 0$ in Theorem 6.2.

Dimension 1: We start with a simple computation which shows that, in dimension $d = 1$, the series $\sum I_n$ diverge. Indeed,

$$1 = \sum_{z:z=0[\mathrm{mod}2],|z|\leq 2t} [P_{t-1}^{\beta,\omega}]^{\otimes 2}(S_t = \widetilde{S}_t + z) \leq (2t + 1)I_t .$$

Hence, $I_t \geq 1/(2t + 1)$ and $\sum_t I_t = \infty$, which shows that $W_\infty = 0$ when $d = 1$. This is an easy proof that $W_\infty = 0$ when $d = 1$.

Dimension 2: We prove the theorem by contradiction. Assume that $W_\infty > 0$ almost surely. Consider on the path space the event

$$A_n = \{|S_n^{(1)}| \leq K\sqrt{n \ln n}, |S_n^{(2)}| \leq K\sqrt{n \ln n}\}$$

where the two coordinates of S_n are smaller in absolute value than $K\sqrt{n \ln n}$. Let

$$X_n = P(e^{\beta H_{n-1}-(n-1)\lambda(\beta)}; A_n^c) .$$

By Markov inequality, for large n,

$$\mathbb{P}\left(X_n \geq e^{-\frac{K^2}{4}\ln n}\right) \leq e^{\frac{K^2}{4}\ln n}\mathbb{P}(X_n)$$

$$= e^{\frac{K^2}{4}\ln n}P(A_n^c)$$

$$\leq 4e^{-\frac{K^2}{4}\ln n}.$$

In the last line we have used Chernov's bound for the random walk in the following way:

$$P(\pm S_n^{(1)} > K\sqrt{n\ln n}) \leq \exp\{-n\gamma^*(K\sqrt{n\ln n})\},$$

with γ^* the convex dual of γ,

$$\gamma(u) := \ln P(e^{uS_n^{(1)}}) = \ln\frac{1 + \cosh u}{2} \leq \ln\frac{1 + e^{u^2/2}}{2} \leq u^2/2,$$

implying that $\gamma^*(v) = \sup_u(uv - \gamma(u)) \geq v^2/2$. Taking $K > 2$, we get $X_n \to 0$ $\mathbb{P}$-almost surely by Borel Cantelli lemma. Then,

$$Y_n := P_{n-1}^{\beta,\omega}(A_n^c) \longrightarrow \frac{0}{W_\infty} = 0 \quad \mathbb{P} - a.s..$$

Hence, denoting by $\mathscr{C}(n, K)$ the cube $\mathscr{C}(n, K) = [-K\sqrt{n\ln n}, K\sqrt{n\ln n}]^2$,

$$(1 - Y_n)^2 = \sum_{x,y \in \mathscr{C}(n,K)} [P_{n-1}^{\beta,\omega}]^{\otimes 2}(S_n = x, \widetilde{S}_n = y)$$

$$\leq \sum_{z \in \mathscr{C}(n,2K)} [P_{n-1}^{\beta,\omega}]^{\otimes 2}(S_n = \widetilde{S}_n + z)$$

$$\leq (4K\sqrt{n\ln n})^2 I_n$$

Therefore, $\mathbb{P}$-a.s., we have $I_n \geq 1/(17K^2 n\ln n)$ ultimately, so $\sum_n I_n = \infty$, contradicting $W_\infty > 0$. This ends the proof. $\qquad\square$

6.2.2 Fractional Moments Estimates

The previous estimates are overly simple, they do not lead to an exponential decay. Building on Derrida's ideas, Lacoin [157] introduced a method of change of measure for the environment to estimate fractional moments ($\mathbb{P}[W_n^\alpha]$ for $\alpha \in (0, 1)$).

It is sharp enough to get $p(\beta) < \lambda(\beta)$ and precise estimates. We summarize successive results from [29, 157, 184] for the discrete model. See also [31] for the parabolic Anderson model.

Theorem 6.3 ([29, 157, 184]) *Assume $\omega(t, x)$ has mean 0 and variance 1.*
For $d = 1$, as $\beta \searrow 0$,

$$\lambda(\beta) - p(\beta) \asymp \beta^4 .$$

For $d = 2$, as $\beta \searrow 0$,

$$\lambda(\beta) - p(\beta) = \exp\left\{-\pi\beta^{-2}(1 + o(1))\right\} .$$

The conjecture in dimension $d = 1$ is that

$$\lambda(\beta) - p(\beta) \sim \frac{1}{24}\beta^4 , \qquad \beta \searrow 0. \tag{6.5}$$

6.3   Simulations of the Localized Phase

We report simulation experiments by Vu-Lan Nguyen from his Ph.D. thesis [189]. The model is the Log-Gamma polymer defined by (7.1)–(7.2) in Chap. 6, with different values of the parameter $\mu = 1$ or 100, and in dimension $d = 1$. The parameter μ is analogous to the inverse temperature $1/\beta$, and the above values respectively correspond to "extremely strong" disorder and "strong but not so strong" disorder. Since the dimension is $d = 1$ we know from Theorem 6.2 that the polymer is localized. We stress that there are no boundary conditions in the simulation.

As a function of time n, we plot in Figs. 6.1, 6.2, 6.3, 6.4, 6.5, and 6.6 the following characteristics of the polymer measure.

1. The mass of the favourite endpoint, J_n from (5.23);
2. The location of the favourite endpoint (after 45° rotation to bring the polymer in the first quadrant);
3. A map of the endpoint density in the full first quadrant.

We can observe the following.

- In the case $\mu = 1$ the localization is strong. Most of the time, the mass of the favourite endpoint is larger than 0.3. There are just a few locations where the mass is not small, and there is a sharp corridor where the polymer tends to lie. From time to time, the favourite endpoint jumps (as n is increased) to a distant location.

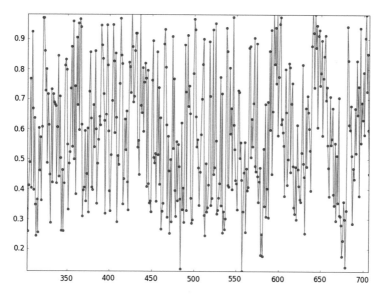

Fig. 6.1 Mass of the favourite endpoint, $\mu = 1$: the mass fluctuates strongly between 0 and 1, taking rather large values from times to times

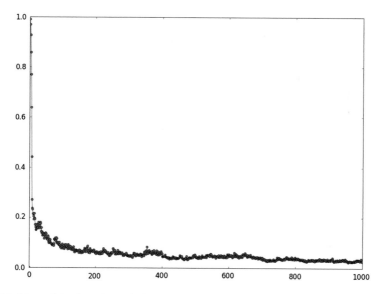

Fig. 6.2 Mass of the favourite endpoint, $\mu = 100$: the mass decreases rapidly in the first steps, but then stabilizes. Theoretical results predict that it has a strictly positive limit in law (a random limit)

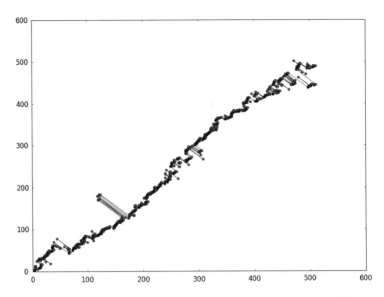

Fig. 6.3 Location of the favourite endpoint, $\mu = 1$: It sticks to an "optimal path" for some time; jumps occur at times when the evolution turns the previously optimal path into suboptimal. Jump size is large at low temperature—corresponding to this case. Back-and-forth oscillations appear

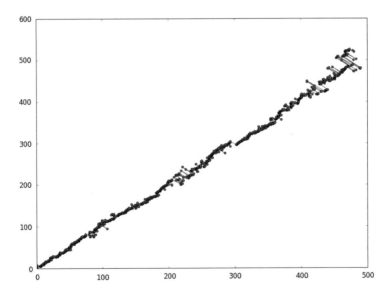

Fig. 6.4 Location of the favourite endpoint, $\mu = 100$: there are fewer jumps, and their size become smaller. The path almost looks like a simple random walk

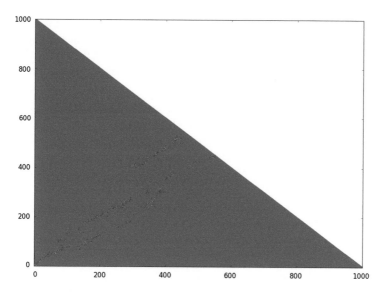

Fig. 6.5 Density profile, $\mu = 1$. Color code: the mass decreases from *red* to *yellow* (mass of the order of the maximum), *green* and *grey* (negligible). Corresponding to very low temperature, the path is pinned in narrow corridors

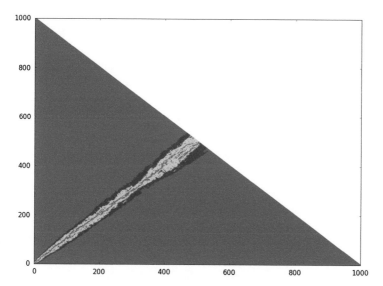

Fig. 6.6 Density profile, $\mu = 100$. Color code: the mass decreases from *red* to *yellow* (mass of the order of the maximum), *green* and *grey* (negligible). The path is pinned on a wider corridor

- If $\mu = 100$, the mass of the favourite endpoint decays rapidly but then stabilizes to a small but positive value. Around the favourite endpoint the mass decays slowly, resulting in a wide corridor. Transverse jumps of the favourite location occur, but at a smaller scale. Though the system is at strong disorder, the localization is not as sharp as above.

I will not resist to the pleasure to quote a sentence from physics Nobel prize Philip Anderson, in his Nobel Lecture on 8 December 1977:

```
It [Localization] has yet to receive adequate
mathematical treatment, and one has to resort to
the indignity of numerical simulations
to settle even the simplest questions about it.
```

Fortunately, much has been done since that time ! In the next section and next chapter we will give an analytical account on localization.

6.4 Localization for Heavy-Tails Environment

We consider now environments ω which are heavy-tailed: in particular the exponential moments are infinite, and the previous approach does not apply. Localization mechanism depends on how heavy is the tail. When $\omega(0, 0)$ has infinite exponential moments but finite $(d + 1)$-th moment—which is heavy tail case, but not extremely heavy—using that the free energy $p(\beta)$ is finite in contrast with an infinite $\lambda(\beta)$, Vargas [235] shows that the law of the endpoint of the polymer has "macroscopic atoms": this is a localization statement as above, from a quite different proof.

In this section, we develop a case of extremely heavy tails. Localization is of a different nature here, it is much stronger and simpler, so it can be completely analysed.

6.4.1 The Big Picture

Auffinger and Louidor in [18] address the case where ω are i.i.d. but their marginal distribution has a (right) tail which is heavy enough to fall outside the KPZ universality class. Inspired by [133] for Last Passage Percolation (LPP), the tail of the environmental distribution decays as an inverse power $\alpha \in (0, 2)$,

$$\mathbb{P}(\omega(t, x) > u) \simeq u^{-\alpha}.$$

Assume for convenience that this distribution has no atoms[1] and that $d = 1$. One can see that for any (fixed) positive finite temperature, the polymer is localized to

---

[1] This will ensure uniqueness of maximizers.

the path along which the energy is minimal, i.e. thermal fluctuations are negligible: this is because entropy is of smaller order compared to energy in this case, and the scaling limit of the Gibbs measure reduces to that of the geodesics given in [133]. Consequently, it is natural to scale temperature $\beta = \beta_n$ with n and indeed if

$$\beta_n \simeq \mathfrak{b} n^{1-2/\alpha},$$

the system exhibits a non-trivial interplay between energy and entropy. In particular, to observe non-trivial paths, the occurrence of large values of ω requires to consider vanishing $\beta = \beta_n$.

Under this scaling, the authors in [18] observe that the environment can be so big at a few points that it can be reduced in the limit $n \to \infty$ to its *extremal process*. A coarse-grained picture is given on the space of curves by an energy taking into account only the largest values of the medium, and an entropy coming from Cramér theory [91]. The polymer chain of length n is localized, with overwhelming probability as n increases, to a cylindrical region of diameter $o(n)$ around a random optimal directed curve. This curve optimally balances entropy and energy at the coarse-grained level, and is at distance of order n. Zero (resp. infinite) temperature behavior[2] is recovered if β_n grows faster (resp. slower) than above.

Weak limit for the distribution of the optimal curve under linear scaling can be constructed as the global solution to a variational problem on the space of curves, see Theorem 6.4 and (6.10) below. The functional being maximized is random, it can be viewed as assigning to a curve the difference between its entropy gain and energy cost under a proper (random) limit environment.

These limiting distributions form a two parameters family of measures on curves $\mathbb{M}_{\alpha,\mathfrak{b}}$, $\alpha \in (0,2)$, $\mathfrak{b} \in [0,\infty]$, including as a special case, the scaling limit of the ground-state path $\mathbb{M}_{\alpha,\infty}$ which was studied in [133].

Here, the transversal fluctuations of the polymer's path are of order n—quite different from the light tail case. This happens for all $\mathfrak{b} > 0$. Nevertheless, as shown in Proposition 2.5 of [18], when α is small enough there exists a random variable $\mathfrak{b}_c$ a.s. positive (but arbitrarily small with positive probability) such that if $\mathfrak{b} < \mathfrak{b}_c$ then the polymer localizes around the x-axis, thereby exhibiting an infinite-temperature behavior (macroscopically looks like a rescaled random walk). This can be viewed as a quenched phase transition with a random threshold value.

As a side remark, we observe that, although we only treat the $1+1$ dimensional case here, the problem is morally the same for all dimensions, with minor changes due to the different geometry. In particular, for any d, the right normalization is $\beta_n = \mathfrak{b} n^{1-(1+d)/\alpha}L(n)$ and the limit curves live inside a $d+1$ regular polyhedron.

---

[2]The reader should recall at this point that we are considering a model where the temperature is not kept constant, but rescaled with n.

6.4.2 The Results

Assume that $d = 1$, and that the tail of the distribution of ω is regularly varying with index $\alpha \in (0, 2)$, namely:

$$\mathbb{P}(\omega(0,0) > u) = u^{-\alpha}L(u). \tag{6.6}$$

where $L(u)$ is a slowly varying function (i.e., $L(au)/L(u) \to 1$ as $u \to \infty$ for all $a > 0$). Let

$$\mathscr{D} = \{(s, y) \in [0, 1] \times [-1, 1] : |y| \le s\}.$$

It is a standard fact (for instance, Sect. 1.1 in [201]) that such distributions are in the max-domain of attraction of the Frechet distribution. Denote by $(U_n^i, Z_n^i)_i$ the value and location in $\mathscr{D}$ of the non-ascending order statistics[3] of of $(\omega(t,x) : (t,x) \in (n\mathscr{D}) \bigcap (\mathbb{N} \times \mathbb{Z}), t \leftrightarrow x)$, recall (3.34). Then, there exist $b_n > 0$ such that for all $k \ge 1$, as $n \to \infty$,

$$\left((b_n^{-1}U_n^i, n^{-1}Z_n^i)\right)_{i=1}^k \xrightarrow{\text{law}} \left((U^i, Z^i)\right)_{i=1}^k, \qquad \forall k, \tag{6.7}$$

where the limit is non-degenerate. The constants b_n can be written as

$$b_n = n^{2/\alpha}L_0(n)$$

where L_0 is another slowly varying function. The random set $((U^i, Z^i); i \ge 1)$ is a realization of a Poisson point process with density given by

$$v(dudsdr) = \frac{\alpha}{u^{1+\alpha}}dudsdr \quad \text{on} \quad (0, +\infty) \times \mathscr{D}.$$

It is a standard fact that $(Z^i)_i$ and $(U^i)_i$ are independent of each other, with

$$Z^i(i \ge 1) \text{ i.i.d. uniform on } \mathscr{D}, \quad \text{and} \quad U^i = (\mathscr{E}_1 + \ldots + \mathscr{E}_i)^{-1/\alpha},$$

with an i.i.d. sequence $(\mathscr{E}_i)_i$ exponentially distributed with mean 1.

Definition 6.1 (Coarse-Grained Entropy) Let us consider the functional space $\mathscr{L}^0 = \{\mathbf{x} : [0; 1] \to \mathbb{R}, 1 - \text{Lipschitz}; \mathbf{x}(0) = 0\}$, equipped with L^∞-norm. For a curve $\mathbf{x} \in \mathscr{L}^0$ its entropy is

$$E(\mathbf{x}) = \int_0^1 e(\dot{\mathbf{x}}(t))dt,$$

---

[3] $U_n^1 > U_n^2 > \ldots$ are the values of ω in the quadrant $\{(t, x) \in (n\mathscr{D}) \bigcap (\mathbb{N} \times \mathbb{Z}), t \leftrightarrow x\}$ in decreasing order, and the $Z_n^i \in [\![1, n]\!] \times \mathbb{Z}$ are the corresponding locations in the quadrant.

where

$$e(x) = \frac{1}{2}[(1+x)\ln(1+x) + (1-x)\ln(1-x)].$$

Then, e is the ± 1-Bernoulli entropy and $E(\cdot)$ is the rate function in the large deviations principle [91] for the sequence of uniform measures on $\mathcal{L}_n^0$, the set of $1/n$-scaled trajectories of a simple random walk. Let us express this statement in a loose way,

$$P\left(n^{-1}S(n\cdot) \simeq \gamma(\cdot)\right) \simeq \exp\{-nE(\gamma)\}. \tag{6.8}$$

A non-trivial behavior is obtained by taking β_n of order n/b_n: Let $\beta_n \geq 0$ with

$$\lim_n n^{-1}b_n\beta_n = \mathfrak{b} \in [0, \infty]. \tag{6.9}$$

We stress that we cover the case $\mathfrak{b} = \infty$. Defining, for $\gamma \in \mathcal{L}^0$,

$$\mathcal{H}_n^\omega(\gamma) = \sum_{i=1}^n \omega(i, [n\gamma(i/n)]),$$

the optimal curve $\gamma_{n,\beta}^*$ is the solution of the following variational problem:

$$\gamma_{n,\beta}^* = \arg\max_{\gamma \in \mathcal{L}^0}\{\beta \mathcal{H}_n^\omega(\gamma) - nE(\gamma)\} \tag{6.10}$$

which exists almost surely. A first results is the concentration of the polymer measure around the optimal curve.

Theorem 6.4 (Theorem 2.1 in [18]) *For all positive ε, δ, and $\mathfrak{b} \in [0, \infty]$, $\exists v > 0$ s.t.*

$$P_n^{\beta_n,\omega}\left(\|n^{-1}S(n\cdot) - \gamma_{n,\beta_n}^*(\cdot)\|_\infty > \delta\right) \leq \begin{cases} e^{-nv} & \text{if } \mathfrak{b} < \infty, \\ e^{-b_n\beta_n v} & \text{if } \mathfrak{b} = \infty. \end{cases}$$

with $\mathbb{P}$-probability greater than $1 - \varepsilon$ for large n.

Define the 'limit environment' as

$$\pi_\infty = \sum_i U^i \delta_{Z^i}.$$

While π_∞ is a measure with infinite mass for $\alpha \geq 1$, $\pi_\infty(\gamma)$ is finite with probability 1 for all path γ in $\mathcal{L}^0$. This follows from Theorem 2.1 in [133]. In view of (6.7), $\pi_\infty(\gamma)$ is a reasonable candidate for a limit of $b_n^{-1}\mathcal{H}_n^\omega(\gamma)$, but a precise statement requires some care. Analogously to (6.10), define another optimal curve $\hat{\gamma}_{\infty,\mathfrak{b}}$ as the solution of the variational problem

$$\hat{\gamma}_{\infty,\mathfrak{b}} = \arg\max_{\gamma\in\mathscr{L}^0}\{\pi_\infty(\gamma) - \mathfrak{b}^{-1}E(\gamma)\}$$

(To take care of the case $\mathfrak{b} = \infty$ we have divided by $\mathfrak{b}$.) Again, a maximizer always exists and it is unique, almost surely. In Theorem 2.2 of [18] it is proved that the rescaled optimal path for finite n converge in law to the optimal path of the limiting variational problem:

$$\gamma^*_{n,\beta_n} \xrightarrow{\text{law}} \hat{\gamma}_{\infty,\mathfrak{b}} .$$

Now, define the random variable

$$\mathfrak{b}_c = \inf\left\{\mathfrak{b} \geq 0 : \max_{\gamma\in\mathscr{L}^0}\{\mathfrak{b}\pi_\infty(\gamma) - E(\gamma)\} > 0\right\}.$$

Theorem 6.5 (Critical Temperature, [18, 231]) .

(i) $\hat{\gamma}_{\infty,\mathfrak{b}} = 0$ if $\mathfrak{b} < \mathfrak{b}_c$, and $\gamma_{\infty,\mathfrak{b}} \neq 0$ if $\mathfrak{b} > \mathfrak{b}_c$.
(ii) For $\alpha < 1/2$, $\mathfrak{b}_c > 0$ a.s.
(iii) For $\alpha \in [1/2, 2)$, $\mathfrak{b}_c = 0$ a.s.

Hence, there exists a critical random threshold $\mathfrak{b}_c$ such that if $\mathfrak{b} < \mathfrak{b}_c$, the path stays close to the straight path between (0,0) and (1,0), while if $\mathfrak{b} > \mathfrak{b}_c$, then it is at distance $O(n)$ from the origin.

The theorem was proved for $\mathfrak{b} \in (0, 1/3) \cup (1/2, 2)$ in [18], leaving open the case [1/3,1/2). The gap was recently bridged by Torri [231].

Remark 6.1

1. It is conjectured in [18] that for any finite $\mathfrak{b}$ the curve $\hat{\gamma}_{\infty,\mathfrak{b}}$ is given by a simple path obtained by a linear interpolation of a finite (random) number of points, but this is still an open question.
2. Let $\mathbb{M}_{\alpha,\mathfrak{b}} = \mathrm{Law}(\hat{\gamma}_{\infty,\mathfrak{b}})$. Then, it is shown in Proposition 2.4 of [18] that

$$\mathfrak{b}^1 \neq \mathfrak{b}^2 \Rightarrow \mathbb{M}_{\alpha,\mathfrak{b}^1} \neq \mathbb{M}_{\alpha,\mathfrak{b}^2}$$

Precisely, for $0 \leq \mathfrak{b}^1 < \mathfrak{b}^2 \leq \infty$, the r.v.'s $E(\hat{\gamma}_{\infty,\mathfrak{b}^1}) \leq_{\text{st}} E(\hat{\gamma}_{\infty,\mathfrak{b}^2})$ are not equal in law, implying the last claim.
3. Torri [231] provides a coupling of the polymer with finite n together with the limiting path measure with environment π_∞.

4. The free energy is random,

$$\lim_{n\to\infty} n^{-1} \ln Z_{n,\beta_n} \overset{\text{law}}{=} \max_{\gamma\in\mathscr{L}^0} \{\mathfrak{b}\pi_\infty(\gamma) - E(\gamma)\},$$

it depends on the environment via the leading order statistics.

6.4.3 Characteristic Exponents and Flory Argument

Recall the roughness and volume exponents introduced in (1.6). From the above results we derive their values in the case $\alpha \in (0, 2)$,

$$\chi^\perp = 1, \quad \chi^\| = 1,$$

and then are equal to those conjectured for $\alpha = 2$, as it is claimed in the following remark:

Remark 6.2 (Heuristics and the Flory Argument) Other values of α are studied in [126], by a Flory argument and numerical simulations. Consider the model with (6.6) and fixed temperature inverse $\beta > 0$—since β_n in (6.9) diverges for $\alpha > 2$. We have:

- For $\alpha \in [2, 5)$, the roughness exponent and the volume exponent are

$$\chi^\perp = \frac{1 + \alpha}{2\alpha - 1}, \quad \chi^\| = \frac{3}{2\alpha - 1}; \tag{6.11}$$

- For $\alpha > 5$, the roughness exponent is $\chi^\perp = 2/3$ and the volume exponent is $\chi^\| = 1/3$, as in the light tail case in space dimension 1.

We recall the so-called **Flory argument**. It applies when the environment is dominated by the largest values of the environment ω in the effective volume. Then, it can be reduced to its extremal field, and the polymer takes a greedy strategy. Denote by r the transverse spread of a typical excursion of the polymer with time horizon n.

- From the theory of extreme statistics of heavy-tailed distributions, we know the largest environment ω in the volume rn available to the polymer, has order of magnitude of $(rn)^{1/\alpha}$. Visiting such a location brings a Boltzmann weight of order roughly

$$\exp\{C(rn)^{1/\alpha}\}.$$

- On the other hand, for the random walk the deformation cost is entropic and, provided that $r \ll n$, follows a scaling similar to that of an elastic energy as r^2/t,

$$P(\max_{0 \leq t \leq n]} |S_t| \geq r) \simeq \exp\{-r^2/(2n)\}$$

The optimal balance is obtained by making the above two exponent equal, leading to the value of $\chi^{\perp}$ in (6.11). The second one follows from the scaling relation $\chi^{\|} = 2\chi^{\perp} - 1$.

The predictions in (6.11) are clearly wrong for $\alpha < 2$ since $\chi^{\perp}$ cannot be larger than 1—so it must be equal to 1 and then the space has to scale like time in this range of values—and the predictions are also wrong for $\alpha > 5$ since at that value the behavior $\chi^{\perp} = 2/3$ of light tails is attained—and from then on, the collective strategy will dominate the greedy strategy.

Chapter 7
Log-Gamma Polymer Model

Recently, significant efforts have been focused on planar polymer models—i.e. $(1 + 1)$-dimensional—which are integrable. In the line of specific first passage percolation models and interacting particle systems, a few explicitly solvable models were discovered, allowing for detailed descriptions of new scaling limits and statistics characteristic of the KPZ universality class.

In this chapter, we study one of these models, introduced by Timo Seppäläinen in his ground-breaking paper [211] in 2012. The model is discrete in time and space and has log-gamma distributed environment. All through the chapter, $\boxed{d=1}$.

7.1 Log-Gamma Model and its Stationary Version

It is convenient to turn the picture of our planar polymer by $45°$ clockwise, see Fig. 7.1, so that the model is in $\mathbb{N}^2$. Also, it is convenient to change notations all through the chapter, and consider the weights

$$Y_{i,j} = e^{\beta\omega(i,j)} \qquad i,j \in \mathbb{N}. \tag{7.1}$$

We will assume $(Y_{i,j}; i,j \geq 1)$ to be i.i.d. with inverse Gamma distribution with parameters $\mu > 0$ and 1, i.e.,

$$Y^{-1} \quad \text{has density} \quad \Gamma(\mu)^{-1}x^{\mu-1}e^{-x} \quad \text{on } (0,+\infty). \tag{7.2}$$

(On the axis we will also have independent weights, but we do not specify the distribution for the moment.) A polymer $\mathbf{x}$ is a nearest-neighbor *up-right* path. In this model, there is no temperature in the strict sense of statistical mechanics (e.g., *no parameter β*), but the parameter $\mu > 0$ plays a similar role, tuning the strength of

© Springer International Publishing AG 2017
F. Comets, *Directed Polymers in Random Environments*, Lecture Notes
in Mathematics 2175, DOI 10.1007/978-3-319-50487-2_7

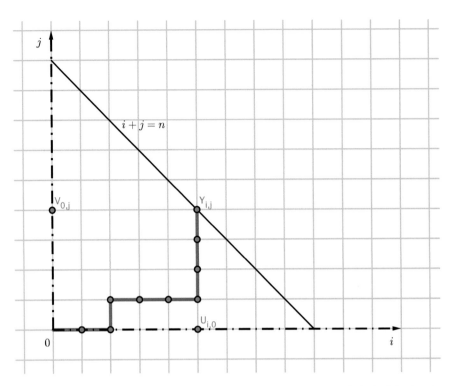

Fig. 7.1 An up-right path of length n with the weights $Y_{i,j}$ on the vertices (i,j). Anticipating the model with boundaries we use the notation U and V for weights on the axis

the disorder. By a plain calculation, we get $\mathbb{P}Y = 1/(\mu - 1)_+$, so the ratio second-to-squared-first moments is

$$\frac{\mathbb{P}(Y^2)}{(\mathbb{P}Y)^2} = \frac{\mu - 1}{\mu - 2}, \quad \mu > 2.$$

Then, small values of μ correspond to large variability of the weights, that is, to large β, or equivalently, to small temperature. We recall that the logarithm $\ln \Gamma(t)$ of the function $\Gamma(t) = \int_0^\infty x^{t-1} e^{-r} dr$ is a convex and infinitely differentiable function on $(0, \infty)$. The derivatives are called polygamma functions, $\Psi_n(t) = (d^{n+1}/dt^{n+1}) \ln \Gamma(t), n = 0, 1, \dots$. For $n \geq 1$, Ψ_n has the sign of $(-1)^{n-1}$ on $(0, \infty)$. The functions Γ, Ψ_0 and Ψ_1 are shown in Fig. 7.2. They come as moments of the log-gamma distribution:

$$\mathbb{P} \ln Y = -\Psi_0(\mu), \quad \text{Var}(\ln Y) = \Psi_1(\mu), \tag{7.3}$$

for Y as above.

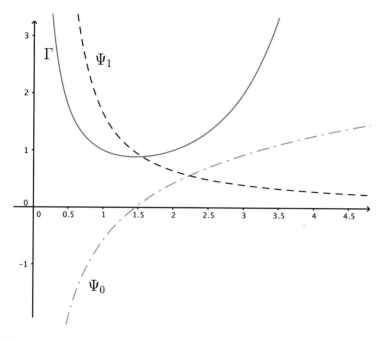

Fig. 7.2 Gamma and polygamma functions: Γ, Ψ_0, Ψ_1

In this chapter we will change notations, and we will not write explicitly the dependence in μ or ω. For each endpoint (i,j) of the path, the point-to-point partition function is

$$Z_{i,j} = \sum_{\mathbf{x} \in \Pi_{i,j}} \prod_{t=1}^{i+j} Y_{x_t},$$

where $\Pi_{i,j}$ denotes the collection of up-right paths $\mathbf{x} = (x_t)_{0 \le t \le i+j}$ in the rectangle $\Lambda_{i,j} = \{0, \ldots, i\} \times \{0, \ldots, j\}$ that start at $(0,0)$ and end at (i,j). We do not normalize by the number $(2d)^n$ of paths. The point-to-line partition function is given by

$$Z_n = \sum_{k=0}^{n} Z_{k,n-k}.$$

The point-to-line polymer measure of a path of length n in the environment $Y = (Y_{i,j}; i,j \ge 0)$ is

$$P_n^Y(\mathbf{x}) = \frac{1}{Z_n} \prod_{t=1}^{n} Y_{x_t}.$$

We now discuss the choice of the environment on the axis. The model we have been considering so far would be to take i.i.d. weight with the same law inverse Gamma(μ, 1). However it is tricky to consider different laws.

Model with boundaries: We assign distinct weight distributions on the boundaries $\mathbb{N} \times \{0\}$, $\{0\} \times \mathbb{N}$ and in the bulk $\mathbb{N}_*^2$. In order to make it clear, we use the symbols U and V for the weights on the horizontal and vertical boundaries:

$$U_{i,0} = Y_{i,0} \text{ and } V_{0,j} = Y_{0,j} \text{ for } i,j \in \mathbb{N}^* := \{1, 2, \ldots\}.$$

Model b.c.(θ): Let $\mu > 0$ be fixed. For $\theta \in (0, \mu)$, we will denote by b.c.(θ) the model with

$\{U_{i,0}, V_{0,j}, Y_{i,j} : i,j \in \mathbb{N}\}$ are independent with distributions

$$U_{i,0}^{-1} \sim \text{Gamma}(\theta, 1), \quad V_{0,j}^{-1} \sim \text{Gamma}(\mu-\theta, 1), \quad Y_{i,j}^{-1} \sim \text{Gamma}(\mu, 1). \tag{7.4}$$

where Gamma(θ, r) distribution has density $\Gamma(\theta)^{-1} r^\theta x^{\theta-1} e^{-rx}$ with $\theta > 0, r > 0$.

In his seminal paper [211], Seppäläinen proved that such boundary conditions make the model stationary, and then exactly solvable. The model is even reversible, as in Burke's theorem: The reader is referred to [101] for Burke's theorem in queueing theory, and to [190] for Brownian queues.

Assume the condition (7.4). Define for $(m, n) \in \mathbb{N}^2$,

$$U_{m,n} = \frac{Z_{m,n}}{Z_{m-1,n}} \text{ and } V_{m,n} = \frac{Z_{m,n}}{Z_{m,n-1}} \; .$$

Then, by (2.3), these obey the recursion

$$U_{m,n} = Y_{m,n} \left(1 + \frac{U_{m,n-1}}{V_{m-1,n}} \right) , \qquad V_{m,n} = Y_{m,n} \left(1 + \frac{V_{m-1,n}}{U_{m,n-1}} \right) . \tag{7.5}$$

The key fact is

Lemma 7.1 *Assume U, V, Y are independent inverse-Gamma distributed, respectively with first parameters $\theta, \mu - \theta, \mu$ and second parameter equal to 1. Define*

$$U' = Y(1 + UV^{-1}) , \quad V' = Y(1 + VU^{-1}) , \quad Y' = (U^{-1} + V^{-1})^{-1} .$$

Then, U', V', Y' are independent with the same laws as U, V, Y.

Proof We start to recall some well-known identities in distribution from the Beta-Gamma algebra.

- For A, B, C independent Gamma with parameters $\theta, \mu - \theta, \mu$, the variables

$$\frac{A}{A+B}, A+B \text{ are independent with laws Beta}(\theta, \mu-\theta) \text{ and Gamma}(\mu).$$

The Beta density is $[\Gamma(\mu)/\Gamma(\theta)\Gamma(\mu - \theta)]x^{\theta-1}(1 - x)^{\mu-\theta-1}$ on $(0, 1)$.
- Conversely, if β, γ are independent with the above Beta and Gamma laws, then $\beta\gamma, (1 - \beta)\gamma$ are independent and Gamma-distributed with respective parameters $\theta, \mu - \theta$.
- For A, B, C as above, $A' = C\frac{A}{A+B}$, $B' = C\frac{B}{A+B}$ are independent and Gamma-distributed with these parameters. Moreover the couple is independent from $C' = A + B$, which has a Gamma(μ) law.

(The first claim is easily checked by change of variable in the joint density, or by Laplace transform computation; the converse holds since the transformation is bijective on the relative domains.)

Taking $A = U^{-1}, B = V^{-1}, C = Y^{-1}$, we obtain the claim in Lemma 7.1. $\qquad \square$

We can associate the U's and V's to edges of the lattice $\mathbb{Z}^2_+$, so that they represent the weight distribution on a horizontal or vertical edge respectively. The boundary weight $U_{i,0}$ is attached to the edge $(i - 1, 0) \to (i, 0)$, $V_{0,j}$ to $(0, j - 1) \to (0, j)$. The weights $U_{m,n}$ and $V_{m,n}$ constructed above are attached to the horizontal and vertical edges ending at (m, n).

Let $\mathbf{e}_1, \mathbf{e}_2$ denote the unit coordinate vectors in $\mathbb{Z}^2$. For an horizontal edge $f = \{y - \mathbf{e}_1, y\}$ we set $T_f = U_y$, and $T_f = V_y$ if $f = \{y - \mathbf{e}_2, y\}$. Let $\mathbf{z} = (z_k)_{k\in\mathbb{Z}}$ be a nearest-neighbor down-right path in $\mathbb{Z}^2_+$, that is, $z_k \in \mathbb{Z}^2_+$ and $z_k - z_{k-1} = \mathbf{e}_1$ or $-\mathbf{e}_2$. Denoting the edges of the path by $f_k = \{z_{k-1}, z_k\}$, we then have

$$T_{f_k} = \begin{cases} U_{z_k}, & \text{if } f_k \text{ is a horizontal edge} \\ V_{z_{k-1}} & \text{if } f_k \text{ is a vertical edge.} \end{cases}$$

See Fig. 7.3.

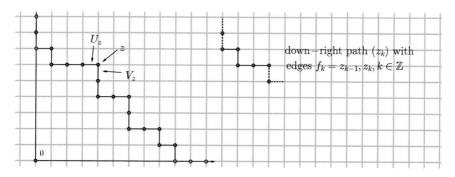

Fig. 7.3 A down-right path. By Theorem 7.1, the weights on the edges of a down-right path are independent in the model with boundaries

Theorem 7.1 (Theorem 3.3 in [211]) *Assume* (7.4). *For any down-right path* $(z_k)_{k \in \mathbb{N}}$ *in* $\mathbb{Z}_+^2$, *the variables* $\{T_{f_k} : k \in \mathbb{Z}\}$ *are mutually independent with marginal distributions*

$$U^{-1} \sim \text{Gamma}(\theta, 1), \quad V^{-1} \sim \text{Gamma}(\mu - \theta, 1).$$

Proof This is proved by induction. If the down-right path coincides with the union of the two semi-axes, then the claim is the assumption (7.4) on the boundary weights. The inductive step consists in adding a "growth corner": suppose that z. goes through the points $(i - 1, j), (i - 1, j - 1)$ and $(i, j - 1)$. Flip the corner over to create a new path ζ' which goes through $(i - 1, j), (i, j)$ and $(i, j - 1)$. Apply Lemma 7.1 with

$$U = U_{i,j-1}, \quad V = V_{i-1,j}, \quad Y = Y_{i,j}, \quad U' = U_{i,j}, \quad V' = V_{i,j},$$

to see that the conclusions hold for the new path.   □

By considering the down-right path along the vertices x with $x \cdot (\mathbf{e}_1 + \mathbf{e}_2) = n$, we deduce the following fact, which will be a fundamental ingredient in the next sections.

Corollary 7.1 *For each n, the variables* $(U_{k,n-k}, V_{k,n-k})_{0 \le k \le n}$ *are independent, and*

$$U_{k,n-k}^{-1} \sim \text{Gamma}(\theta, 1), \quad V_{k,n-k}^{-1} \sim \text{Gamma}(\mu - \theta, 1). \tag{7.6}$$

7.2 Free Energy and Fluctuations for the Stationary Model

From the information on the marginals in Theorem 7.1, we get

$$\mathbb{P}[\ln Z_{m,n}] = m\mathbb{P}(\ln U) + n\mathbb{P}(\ln V) = -m\Psi_0(\theta) - n\Psi_0(\mu - \theta), \tag{7.7}$$

from the values in (7.3). Furthermore,

$$\ln Z_{m,n} = \ln Z_{0,n} + \ln(Z_{m,n}/Z_{0,n}) = \ln Z_{m,0} + \ln(Z_{m,n}/Z_{m,0}) \tag{7.8}$$

is the sum of 2 sums over 2 sets of independent r.v.'s, so the usual law of large numbers for i.i.d.r.v.'s applies to each set, yielding the value of the free energy for the stationary model: recalling (7.3), we have for $s, t \in [0, 1]$,

$$p^{\text{b.c.}}(\theta; s, t) := \lim_{N \to \infty} N^{-1} \ln Z_{[Ns],[Nt]} = -s\Psi_0(\theta) - t\Psi_0(\mu - \theta) \qquad \mathbb{P}-a.s. \tag{7.9}$$

For each choice of the parameters θ, μ, there is a choice of a *characteristic direction*

$$\zeta(\mu, \theta) = \left(\Psi_1(\mu - \theta), \Psi_1(\theta)\right) \in (\mathbb{R}_+^*)^2 \tag{7.10}$$

for the polymer. This is the direction (s, t) such that the minimum of the convex function $\theta' \mapsto p^{b.c.}(\theta'; s, t)$ is achieved at $\theta' = \theta$. With N a scaling parameter that we will take large, assume that the coordinates (m, n) of the endpoint satisfy

$$|m - N\Psi_1(\mu - \theta)| \leq \gamma N^{2/3}, \qquad |n - N\Psi_1(\theta)| \leq \gamma N^{2/3}, \qquad (7.11)$$

for some fixed constant γ. Seppäläinen proves in [211] the following bounds on the free energy variance.

Theorem 7.2 *Assume (7.4) and (7.11). Then there exist constant $0 < C_1 < C_2 < \infty$ such that*

$$C_1 N^{2/3} \leq \text{Var}(\ln Z_{m,n}) \leq C_2 N^{2/3}$$

Thus, in presence of boundary conditions, the fluctuation exponent of free energy in the characteristic direction is $\chi^\| = 1/3$.

The values of the characteristic exponents $\chi^\| = 1/3, \chi^\perp = 2/3$, is believed to be universal for one-dimensional polymers. These values were found by non-rigorous arguments in physics literature [110, 150].

Proof Short sketch of the main steps of the (technical!) proof: It is necessary to understand the (strong) correlation between the two sets of i.i.d.r.v.'s mentioned above. Firstly, let's consider the exit points of a path X from the horizontal [resp., vertical] axis,

$$\zeta_x = \max\{k \geq 0 : X_i = (i, 0) \, \forall i \leq k\}, \quad \zeta_y = \max\{k \geq 0 : X_j = (0, j) \, \forall j \leq k\}.$$

Remark that one of ζ_x, ζ_y is zero, the other one being positive. Introduce the function

$$L(\theta, x) = \int_0^x (\Psi_0(\theta) - \ln y) x^{-\theta} y^{\theta-1} e^{x-y} dy, \qquad \theta, x > 0, \qquad (7.12)$$

which is also equal to

$$L(\theta, x) = -\Gamma(\theta) x^{-\theta} e^x \times \text{Cov}(\ln A, 1_{\{A \leq x\}}), \quad A \sim \text{Gamma}(\theta, 1).$$

One can check that $L > 0$ and $\mathbb{P}L(\theta, A) = \Psi_1(\theta)$. In [211], it is first established that, with $N = n + m$,

$$\text{Var}(\ln Z_{m,n}) = \mathbb{P}P_N^Y \left[\sum_{1 \leq i \leq \zeta_x} L(\theta, Y_{i,0}^{-1}) \Big| x_N = (m, n) \right]$$

$$+ \mathbb{P}P_N^Y \left[\sum_{1 \leq j \leq \zeta_y} L(\mu - \theta, Y_{0,j}^{-1}) \Big| x_N = (m, n) \right].$$

Then, two rough estimates are established:

- The variables L entering the previous formula have contributions of order 1,

$$\mathrm{Var}(\ln Z_{m,n}) \asymp \mathbb{P}P_N^Y \left[\zeta_x + \zeta_y \big| x_N = (m,n) \right] ;$$

- Under the point-to-point polymer measure in the characteristic direction,

$$\zeta_x + \zeta_y \equiv \zeta_x \vee \zeta_y \asymp N^{2/3} \qquad \text{with high probability.} \tag{7.13}$$

Now, Theorem 7.2 follows from these estimates. $\square$

The result calls for a few consequences and comments. First, it states that, when the polymer moves in the characteristic direction as indicated in (7.11), the fluctuation of the free energy has order $N^{1/3}$, so $\chi^{\|} = 1/3$.[1] Since m, n are proportional to N, each of the 2 terms in the sum (7.8) has fluctuations of order $N^{1/2}$, but the 2 are so strongly correlated that the sum fluctuates at a lower order. In the course of the proof, we have seen that the transverse fluctuations of the path is at least of order $N^{2/3}$, see (7.13). At time $\zeta_x + \zeta_y$ the path leaves the axes, and in fact transverse fluctuations will be of the same order in the remaining time. Thus, $\chi^{\perp} = 2/3$.

We end by describing the scenario when the *end point is far from the characteristic direction*. Then, the decomposition (7.8) of $\ln Z_{m,n}$ should be replaced by a sum $\ln Z_{k,l} + \ln(Z_{m,n}/Z_{k,k})$ where (k, l) is the projection on (m, n) on the characteristic direction: then the first term fluctuates at order $N^{1/3}$, though the second one fluctuates at order of square root of the distance to the diagonal. The first fluctuations being negligible in front of the second ones, and the sum is asymptotically normal. For simplicity we only give the case where the horizontal distance is too large:

Assume (7.4), assume $n = [N\Psi_1(\theta)]$ and also

$$\lim_{N\to\infty} N^{-\alpha}(m - N\Psi_1(\mu - \theta)) \in (0, \infty)$$

for some $\alpha \in (2/3, 1]$. Then, as $N \to \infty$,

$$N^{-\alpha/2}\left(\ln Z_{m,n} - \mathbb{P}[\ln Z_{m,n}] \right) \xrightarrow{\text{law}} \mathcal{N}(0, \sigma^2)$$

for some explicit $\sigma \in (0, \infty)$.

---

[1] See (7.35) for a more precise result.

7.3 Model Without Boundaries

It is convenient to introduce new notations, for $1 \leq k \leq m, 1 \leq l \leq n$,

$$Z^{\square}_{(k,l),(m,n)} = \sum_{\mathbf{x}:(k,l)\to(m,n)} \prod_{t=k+l}^{m+n} Y_{x_t},$$

where the sum is over all up-right paths $\mathbf{x} = (x_t)_t$ that start at (k, l) and end at (m, n). Quantities with superscript "$\square$" do not depend on the boundary conditions, they are those of the usual polymer model. On the contrary to $Z^{\square}_{(k,l),(m,n)}$, $Z_{m,n} = Z^{(\theta)}_{m,n}$ depends on the boundary conditions, so we will indicate explicitly what parameter θ is used.

Theorem 7.3 *Assume* (7.1), *i.e., consider the model without boundaries. For $s, t \geq 0$, the following point-to-point free energy exists $\mathbb{P}$-a.s., and is given by the minimizer problem:*

$$p(s, t) \overset{\text{def}}{=} \lim_{N\to\infty} N^{-1} \ln Z^{\square}_{(1,1),(Ns,Nt)} = \inf_{0<\theta<\mu} \{-s\Psi_0(\theta) - t\Psi_0(\mu - \theta)\}$$

Proof First, we anticipate the existence of the a.s.-limit $p(s, t)$, see Theorem 9.1. We enlarge our probability space in order to accommodate environment on the boundary so that (7.4) holds, and we express

$$Z^{(\theta)}_{Ns,Nt} = \sum_{k=1}^{[Ns]} \left(\prod_{i=1}^{k} U_{i,0}\right) Z^{\square}_{(k,1),(Ns,Nt)} + \sum_{k=1}^{[Nt]} \left(\prod_{i=1}^{k} V_{0,i}\right) Z^{\square}_{(1,k),(Ns,Nt)},$$

$$\lim_{N\to\infty} \frac{1}{N} \ln Z^{(\theta)}_{Ns,Nt} = \sup_{0\leq a\leq s} \{-a\Psi_0(\theta) + p(s - a, t)\}$$

$$\vee \sup_{0\leq b\leq t} \{-b\Psi_0(\mu - \theta) + p(s, t - b)\},$$

by coarse-graining and the law of large numbers. Using the value of $p^{\text{b.c.}}$ in (7.9), taking $s = t$ and using the symmetry $p(s, t) = p(t, s)$, we get

$$-t\big(\Psi_0(\theta) + \Psi_0(\mu - \theta)\big) = \sup_{0\leq a\leq t} \{-a\big(\Psi_0(\theta) \wedge \Psi_0(\mu - \theta)\big) + p(t - a, t)\}.$$

We restrict to $\theta \in (0, \mu/2]$, so that $\Psi_0(\theta) \wedge \Psi_0(\mu - \theta) = \Psi_0(\theta)$ by monotonicity of Ψ_0, and we change variables $a = t - s$:

$$-t\Psi_0(\mu - \theta) = \sup_{0\leq s\leq t} \{s\Psi_0(\theta) + p(s, t)\}.$$

We turn this into a convex duality by changing variables $v = \Psi_0(\theta)$:

$$-t\Psi_0\big(\mu - \Psi_0^{-1}(v)\big) = \sup_{0 \le s \le t} \{sv + p(s,t)\}, \quad v \in (-\infty, \Psi_0(\mu/2)]. \quad (7.14)$$

Fix $t > 0$, and anticipate that $p(\cdot, t)$ is concave continuous on $[0, t]$, see Theorem 9.1. Extend $f(s) = -p(s,t)$ to a lower semicontinuous convex function on $\mathbb{R}$ by $f(s) = \infty$ for $s \notin [0, t]$. Then, (7.14) writes

$$f^*(v) = -t\Psi_0\big(\mu - \Psi_0^{-1}(v)\big), \quad v \in (-\infty, \Psi_0(\mu/2)].$$

and we can invert by taking double convex duality,

$$f(s) = \sup_{v \in \mathbb{R}} \{sv - f^*(v)\},$$

and we restrict now the domain of optimization. From its explicit expression we differentiate f^* and check that $(f^*)'(-\infty) = 0, (f^*)'(\Psi_0(\mu/2)) = t$. Then, the above supremum can be restricted,

$$f(s) = \sup_{v \in (-\infty, \Psi_0(\mu/2)]} \{sv - f^*(v)\}.$$

Undoing the change of variable, we get for $s \in [0, t]$,

$$-p(s,t) = \sup_{\theta \in (0, \mu/2]} \{s\Psi_0(\theta) + t\Psi_0(\mu - \theta)\},$$

which yields the desired formula in the theorem since $p(s,t) = p(t,s)$. $\qquad\square$

Corollary 7.2 (Free Energy) *Assume* (7.1), *i.e., consider the model without boundaries. Then,*

$$\lim_{n \to \infty} n^{-1} \ln Z_n = -\Psi_0(\mu/2).$$

Proof Anticipating (9.4) in Theorem 9.1, we write

$$
\begin{aligned}
\lim_{n \to \infty} n^{-1} \ln Z_n \;&=\; \sup_{0 \le s \le 1} p(s, 1-s) \\[4pt]
&=\; \sup_{0 \le s \le 1} \inf_{0 < \theta < \mu} \{-s\Psi_0(\theta) - (1-s)\Psi_0(\mu - \theta)\} \\[4pt]
&\overset{\text{minmax th.}}{=}\; \inf_{0 < \theta < \mu} \sup_{0 \le s \le 1} \{-s\Psi_0(\theta) - (1-s)\Psi_0(\mu - \theta)\} \\[4pt]
&=\; \inf_{0 < \theta < \mu} \{-[\Psi_0(\theta) \wedge \Psi_0(\mu - \theta)]\}
\end{aligned}
$$

yielding the desired result. We have used the min-max theorem,[2] which requires to restrict to a compact domain of θ, but this extra constraint can be imposed without changing the value, as it is easily checked. □

Remark 7.1 (Wandering Exponent)
   In [211] fluctuations of the paths are also studied, establishing that the exponent for transverse displacement of the path is $\chi^{\perp} = 2/3$.
   Two cases are considered: (i) point-to-line polymer measures in the presence of boundaries (Theorem 2.3), (ii) point-to-point polymer measures without boundaries (Theorem 2.5). The middle point of a polymer of length N is typically at distance of order $N^{2/3}$ from the characteristic direction in the first case, and from the straight line between endpoints in the second case.

7.4 Localization in the Log-Gamma Polymer with Boundaries

With general techniques and in a general context, we have revealed a localization phenomenon at strong disorder. Due to the complete generality, the results were qualitative. In this section, we consider the log-gamma model, taking advantage of its solvability to analyze the mechanism of localization and obtain an explicit description. In [71] we obtain an explicit limit description of the endpoint distribution under the quenched measure,

$$P_n^Y\{x_n = (k, n - k)\} = \frac{Z_{k,n-k}}{Z_n}, \quad k = 0, \ldots, n.$$

For each n, denote by

$$l_n = \arg\max\{Z_{k,n-k}; 0 \leq k \leq n\}, \tag{7.15}$$

the location maximizing the above probability, i.e., the so-called "favourite end-point".

Theorem 7.4 ([71]) *Consider the model b.c.(θ) with $\theta \in (0, \mu)$. Define the end-point distribution $\tilde{\xi}^{(n)}$ centered around its mode, by*

$$\tilde{\xi}^{(n)} = (\tilde{\xi}_k^{(n)}; k \in \mathbb{Z}), \quad with \ \tilde{\xi}_k^{(n)} = P_n^Y\{x_n = (l_n + k, n - l_n - k)\}. \tag{7.16}$$

---

[2]Here is a simple version (e.g., [91, Exercise 2.2.38]). Let $X \subset \mathbb{R}^n$ and $Y \subset \mathbb{R}^m$ be compact convex sets. If $g : X \times Y \to \mathbb{R}$ is a continuous function that is convex-concave, i.e. $g(\cdot, y) : X \to \mathbb{R}$ is convex for fixed y, and $g(x, \cdot) : Y \to \mathbb{R}$ is concave for fixed x. Then we have that

$$\min_{x \in X} \max_{y \in Y} g(x, y) = \max_{y \in Y} \min_{x \in X} g(x, y).$$

Thus, $\tilde{\xi}^{(n)}$ is a random element of the set $\mathcal{M}_1$ of probability measures on $\mathbb{Z}$. Then, as $n \to \infty$, we have convergence in law

$$\tilde{\xi}^{(n)} \xrightarrow{\mathcal{L}} \xi \qquad \text{in the space } (\mathcal{M}_1, \|\cdot\|_{TV}), \tag{7.17}$$

where $\|\mu - v\|_{TV} = \sum_k |\mu(k) - v(k)|$ is the total variation distance. The limit ξ is defined in (7.23).

Proof Fix n and define for each $1 \le k \le n$ the random variable X_k^n

$$X_k^n = -\log(\frac{Z_{k,n-k}}{Z_{k-1,n-k+1}}) = -\log(\frac{U_{k,n-k}}{V_{k-1,n-k+1}}),$$

and $X_0^n = 0$. By Corollary 7.1, for each n, $(X_k^n)_{1 \le k \le n}$ are i.i.d random variables, and satisfy

$$\frac{Z_{k,n-k}}{Z_{0,n}} = \exp(-\sum_{i=0}^k X_i^n). \tag{7.18}$$

Defining $\mathfrak{S}_k^n = \sum_{i=1}^k X_i^n$, for $0 \le k \le n$, we will express the mass at point $(k, n-k)$ as a function of $\mathfrak{S}^n$,

$$P_n^Y\{x_n = (k, n-k)\} = \frac{Z_{k,n-k}}{\sum_{i=0}^n Z_{i,n-i}} = \frac{1}{\sum_{i=0}^n \exp(-(\mathfrak{S}_i^n - \mathfrak{S}_k^n))}$$

From (7.18), the favourite point l_n defined in (7.15) is also the minimum of the random walk,

$$l_n = \arg\min\{\mathfrak{S}_k^n; 0 \le k \le n\}. \tag{7.19}$$

Since we are only interested in the law of $P_n^Y\{x_n = (k, n-k)\}$, in order to simplify the notation, we consider a single set of i.i.d random variables $(X_k)_{k \in \mathbb{Z}_+}$, with the same distribution under $\mathbb{P}$ as $\log(U/V)$, where U and V are independent with the same distribution as in (7.6). The associated random walk is given by

$$\mathfrak{S}_n = \sum_{i=1}^n X_i, \tag{7.20}$$

and we define

$$\xi_k^n = \frac{1}{\sum_{i=0}^n \exp(-(\mathfrak{S}_i - \mathfrak{S}_k))}.$$

Then one can check that for every n:

$$(\xi_k^n)_{0 \leq k \leq n} \overset{\mathcal{L}}{=} \left(P_n^Y \{ x_n = (k, n-k) \} \right)_{0 \leq k \leq n},$$

where $\overset{\mathcal{L}}{=}$ means equality in law. Then instead of considering for each n a new set of i.i.d random variables to calculate $\tilde{\xi}_k^{(n)}$, we just need the n first steps of the random walk $\mathfrak{S}_n$ to compute the law of ξ_k^n. Hence Theorem 7.4 can be reformulated as follows:

$$\{ \xi_{\ell_n+k}^n \}_{k \in \mathbb{Z}} \overset{\mathcal{L}}{\longrightarrow} \{ \xi_k \}_{k \in \mathbb{Z}}, \quad \text{in the } \ell_1 - \text{norm}, \tag{7.21}$$

with

$$\ell_n = \arg \min_{k \leq n} \mathfrak{S}_k. \tag{7.22}$$

Hence, the localization of the polymer around the favourite point directly relates to the problem of splitting a random walk at its local minima. This coupling is also the main tool to study the recurrent random walk in random environment [114] in one dimension. In the literature, it was proved by Williams [238], Bertoin [33], Bertoin and Doney [34], that if the random walk is split at its local minimum, the two new processes will converge in law to certain limits which are related to a process called the random walk conditioned to stay positive/negative. The mechanism is reminiscent of Sinaï localization, i.e., the localization in the main valley of the one-dimensional random walk in random environment in the recurrent case, discovered by Sinaï [215] and studied by Golosov [121].

We first give an informal definition, which is different according to $\theta = \mu/2$ or not.

- Case $\boxed{\theta = \mu/2}$.

Let $(\mathfrak{S}_k^{\uparrow}, k \geq 0)$, $(\mathfrak{S}_k^{\downarrow}, k \geq 0)$ be two independent processes, with the first one distributed as the random walk $\mathfrak{S}$ conditioned to be non-negative (forever), and the second one distributed as the random walk $\mathfrak{S}$ conditioned to be positive (for positive k). Then,

$$\xi_k = \begin{cases} \dfrac{\exp(-\mathfrak{S}_k^{\uparrow})}{1 + \displaystyle\sum_{i=1}^{\infty} \exp(-\mathfrak{S}_i^{\uparrow}) + \sum_{i=1}^{\infty} \exp(-\mathfrak{S}_i^{\downarrow})}, & \text{if } k \geq 0, \\[2em] \dfrac{\exp(-\mathfrak{S}_k^{\downarrow})}{1 + \displaystyle\sum_{i=1}^{\infty} \exp(-\mathfrak{S}_i^{\uparrow}) + \sum_{i=1}^{\infty} \exp(-\mathfrak{S}_i^{\downarrow})}, & \text{if } k < 0. \end{cases} \tag{7.23}$$

Since we condition by a negligible event, the proper definition requires some care, it relies on Doob's h-transform. It is well known that the limit

$$V(x) = \lim_{n\to\infty} P_x[\mathfrak{S}_k \geq 0, \ k = 1, 2, \ldots n]/P_0[\mathfrak{S}_k \geq 0, \ k = 1, 2, \ldots n]$$

exists and is harmonic for the chain $\mathfrak{S}$. Now we define Doob's h-transform P_x^V of P_x by the function V, i.e., the law of the homogeneous Markov chain on the nonnegative real numbers with transition function

$$p^V(x, y) = \frac{V(y)}{V(x)} p(x, y) 1_{\{y \geq 0\}}, \tag{7.24}$$

where p denotes the transition density of the random walk $\mathfrak{S}$ on $\mathbb{R}$. Denote by $(\mathfrak{S}_k^\uparrow)_{k \geq 0}$ the chain starting from 0,

$$\mathbb{E}(f(\mathfrak{S}_1^\uparrow, \ldots, \mathfrak{S}_k^\uparrow)) = \mathbb{E}_0^V(f(\mathfrak{S})). \tag{7.25}$$

The following result shows that it yields the correct description of the random walk conditioned to stay nonnegative: For a bounded function $f(\mathfrak{S}) = f(\mathfrak{S}_1, \ldots, \mathfrak{S}_k)$,

$$\lim_{n\to\infty} \mathbb{E}_0(f(\mathfrak{S})|\mathfrak{S}_k \geq 0, \ k = 1, 2, \ldots n) = \mathbb{E}(f(\mathfrak{S}^\uparrow)). \tag{7.26}$$

The construction of $\mathfrak{S}^\downarrow$ is similar. The relevance for our purpose is the following convergence:

Lemma 7.2 *For fixed K, we have, as $n \to \infty$:*

$$(\mathfrak{S}_{\ell_n+k} - \mathfrak{S}_{\ell_n})_{1 \leq k \leq K} \xrightarrow{\mathcal{L}} (\mathfrak{S}_k^\uparrow)_{1 \leq k \leq K}, \tag{7.27}$$

$$(\mathfrak{S}_{\ell_n+k} - \mathfrak{S}_{\ell_n})_{-1 \geq k \geq -K} \xrightarrow{\mathcal{L}} (\mathfrak{S}_k^\downarrow)_{1 \leq k \leq K}. \tag{7.28}$$

To control infinite sums we will use a control of the escape of the conditioned random walk:

Lemma 7.3 ([202]) *Fixed $\eta < 1/2$ then :*

$$\liminf_{\delta \to 0} \ _n \mathbb{P}\Big[\inf_{k \leq n}(\mathfrak{S}_k - \delta k^\eta) \geq 0 | \mathfrak{S}_k \geq 0, k = 0, 1 \ldots n\Big] = 1.$$

The reader will find additional details and a proof of these lemmas in [71].

- Case $\boxed{\theta < \mu/2}$.

In this case, $l_n = \mathcal{O}(1)$, but the limit is still given by the formula (7.23), provided that $\mathfrak{S}^\downarrow$ has a lifetime (equal to the time for the walk to reach its absolute minimum), after which it is infinite. $\mathfrak{S}^\uparrow$ is as before, and it is defined in a classical manner. Thus, the concatenated process is simply equal to $\mathfrak{S}$ with a space shift by its minimum value, and time shift by the time to reach the minimum. Since we condition on an event with nonzero probability, the construction of the conditional process is now trivial.

- Case $\boxed{\theta > \mu/2}$.

The last case is similar to the previous one under the change $k \mapsto n - k$.

With all these preambles, the limit ξ given by the formula (7.23) is defined, and we can complete the proof: consider a fixed k, say $k \geq 0$ for definiteness, and write

$$
\begin{aligned}
\xi^n_{\ell_n+k} &= \frac{\exp(-\mathfrak{S}_{\ell_n+k})}{\sum_{i=0}^n \exp(-\mathfrak{S}_i)} \\
&= \frac{\exp(-(\mathfrak{S}_{\ell_n+k} - \mathfrak{S}_{\ell_n}))}{\sum_{i=0}^n \exp(-(\mathfrak{S}_i - \mathfrak{S}_{\ell_n}))} \\
&\xrightarrow{\mathcal{L}} \frac{\exp(-\mathfrak{S}^{\uparrow}_k)}{1 + \sum_{i=1}^{\infty} \exp(-\mathfrak{S}^{\uparrow}_i) + \sum_{i=1}^{\infty} \exp(-\mathfrak{S}^{\downarrow}_i)} = \xi_k ,
\end{aligned}
\tag{7.29}
$$

by Lemmas 7.2 and 7.3. This gives weak convergence to the desired limit. To upgrade weak convergence of the sequence to ℓ_1-norm, it suffices[3] to observe that the limit ξ has norm 1. □

Theorem 7.4 has a number of consequences. In particular, the mass of the favourite point is converging. Recall the definitions (5.7) and (5.23) of I_n, J_n, that we reformulate with the notations of this chapter,

$$
I_n = \sum_{k=0}^n P^Y_n\{x_n = (k, n-k)\}^2 , \qquad J_n = \max_{0 \leq k \leq n} P^Y_n\{x_n = (k, n-k)\},
$$

Corollary 7.3 *Consider the model b.c.(θ) from (7.4). Then, as $n \to \infty$,*

$$
I_n \xrightarrow{\mathcal{L}} \sum_{k \in \mathbb{Z}} \left\{ \frac{\xi_k + \xi_{k+1}}{2} \right\}^2 > 0,
$$

and

$$
J_n \xrightarrow{\mathcal{L}} \max \left\{ \frac{\xi_k + \xi_{k+1}}{2} ; k \in \mathbb{Z} \right\} > 0.
$$

---

[3]For a sequence of probability measures μ_n on $\mathbb{Z}$, pointwise convergence to a probability measure ($\lim_n \mu_n(k) = \mu(k)$ with $\sum_k \mu(k) = 1$) implies convergence in total variation ($\lim_n \|\mu_n - \mu\|_1 = 0$).

Proof This comes directly from continuity. $\qquad\square$

Moreover, we derive that the endpoint density indeed concentrates in a microscopic region, i.e., of size $\mathcal{O}(1)$, around the favourite endpoint.

Corollary 7.4 (Tightness of Polymer Endpoint) *Consider the model b.c.(θ) from (7.4) with $\theta \in (0, \mu)$. For all sequence $K_n \to \infty$, we have*

$$\lim_{n\to\infty} P_n^Y\big[\|x_n - (l_n, n-l_n)\| \geq K_n\big] = 0 \text{ in probability.} \qquad (7.30)$$

Proof From Theorem 7.4, we have for a fixed K,

$$P_n^Y\big[\|x_n - (l_n, n-l_n)\| \geq K\big] \;=\; \sum_{|k|\geq K} \xi_{\ell_n+k}^n$$

$$\xrightarrow{\text{law}} \sum_{|k|\geq K} \xi_k,$$

where the limit goes to 0 a.s. as $K \to \infty$. The claim follows easily. $\qquad\square$

In particular in the equilibrium case $\theta = \mu/2$, $\mathfrak{S}_k$ is a random walk with expectation zero. By Donsker's invariance principle, the random walk has a scaling limit,

$$\left(\frac{1}{\sqrt{n}}\mathfrak{S}_{[nt]}\right)_t \xrightarrow{\mathscr{L}} (W_t)_t . \qquad (7.31)$$

with W a Brownian motion with diffusion coefficient $2\Psi_1(\mu/2)$ (there, $\Psi_1(\theta) = (\log \Gamma)''(\theta)$ is the trigamma function). By consequence, the scaling limit of the favourite endpoint is easy to compute in the present model with boundary conditions.

Theorem 7.5 *Consider the model b.c.(θ) from (7.4).*

(i) *When $\theta = \mu/2$, we have*

$$\frac{l_n}{n} \xrightarrow{\mathscr{L}} \arg\min_{t\in[0,1]} W_t ,$$

where the limit has the arcsine distribution with density $\left[\pi \sqrt{s(1-s)}\right]^{-1}$ on the interval $[0, 1]$.

(ii) *When $\theta < \mu/2$, $n - l_n$ converges in law, so*

$$\frac{l_n}{n} \xrightarrow{\mathbb{P}} 1,$$

though when $\theta > \mu/2$, l_n converges in law, so

$$\frac{l_n}{n} \xrightarrow{\mathbb{P}} 0.$$

In words, the favourite location for the polymer endpoint is random at a macroscopic level in the equilibrium case, and degenerate otherwise. Further, the (doubly random) polymer endpoint x_n has the same asymptotics under Q_n^ω, since, by (7.30), x_n/n and l_n/n are asymptotic as $n \to \infty$. These results disagree with KPZ theory, where the endpoint fluctuates at distance $n^{2/3}$ around the diagonal. A word of explanation is necessary. The difference comes from the boundary conditions. In the equilibrium case $\mu/2 = \theta$ the direction of the endpoint has a maximal dispersion, though in non equilibrium ones it sticks to one of the coordinate axes. With these boundary conditions the limiting shape is linear, thus it does not have the curvature property leading to the KPZ exponents. In the model without boundary conditions—that we leave untouched in this paper—we expect an extra entropy term to come into the play and balance the random walk $\mathfrak{S}_n$ in the potential, a factor being of magnitude n and quadratic around its minimum (which is the diagonal by symmetry), making the localization happen close to the diagonal and with fluctuations of order $n^{2/3}$.

Finally, we derive a large deviation principle for the endpoint distribution:

Theorem 7.6 *Consider the model b.c.(θ) from (7.4).*

(i) Assume $\theta = \mu/2$. In the Skorohod space $D\left([0,1], \mathbb{R}^+\right)$ equipped with Skorohod topology,

$$\left(\frac{-1}{\sqrt{n}} \log Q_n^\omega \{x_n = ([ns], n-[ns])\}\right)_{s \in [0,1]} \xrightarrow{\mathscr{L}} \left(W(s) - \min_{[0,1]} W\right)_{s \in [0,1]}. \tag{7.32}$$

Moreover, for all segment $A \subset \{(s, 1-s); s \in [0,1]\}$ in the first quadrant,

$$\frac{-1}{\sqrt{n}} \log Q_n^\omega (x_n \in nA) \xrightarrow{\mathscr{L}} \inf_A W - \min_{[0,1]} W. \tag{7.33}$$

(ii) Assume $\theta > \mu/2$. Then, as $n \to \infty$,

$$-\frac{1}{n} \log Q_n^\omega \{x_n = ([ns], n-[ns])\} \xrightarrow{\mathbb{P}} s \left|\Psi_0(\theta) - \Psi_0(\mu - \theta)\right|, \tag{7.34}$$

where $\Psi_0(\theta) = (\log \Gamma)'(\theta)$ is the digamma function. Similarly, if $\theta < \mu/2$,.

$$-\frac{1}{n} \log Q_n^\omega \{x_n = ([ns], n-[ns])\} \xrightarrow{\mathbb{P}} (1 - s) \left|\Psi_0(\theta) - \Psi_0(\mu - \theta)\right|.$$

Then, at logarithmic scale, the large deviation probability for the endpoint is of order $\sqrt{n}$ in the equilibrium case, whereas it is of order n otherwise. This is again specific to boundary conditions, since it is shown in [116] for the model without boundaries that the large deviation probabilities have exponential order n with a rate function which vanishes only on the diagonal ($s = 1/2$).

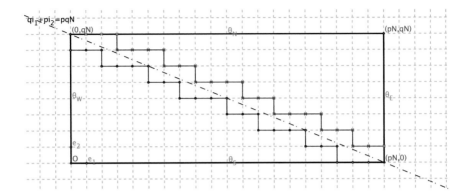

Fig. 7.4 Point-to-point log-gamma polymer with b.c. $\theta_S + \theta_W = \theta_N + \theta_E = \mu$. With high quenched probability, the polymer path crosses the second diagonal (line from $(0, qN)$ to $(pN, 0)$) at a microscopic distance from a "favourite point". The favourite point is random, and distributed at a macroscopic scale as arc-sine law on the second diagonal

Remark 7.2 Analogous results for the middle point of point-to-point polymers with b.c. are given in [71]. To get a flavor, fix two integers $p, q \geq 1$ and consider the log-gamma polymer with end point (pN, qN), and boundary conditions on the four sides of the box, see Fig. 7.4. Assume equilibrium, i.e., the parameters on the South and West sides add up to μ, as well as those on the North and East sides: $\theta_N + \theta_W = \theta_N + \theta_E = \mu$. Then, the crossing of the second diagonal of the box by the polymer will occur at an arc-sine distributed point.

7.5 Final Remarks and Complements

In the line of specific first passage percolation models and interacting particle systems, a few exactly solvable models were discovered in space dimension $d = 1$, allowing for detailed descriptions of new scaling limits and statistics characteristic of the KPZ universality class. Exactly solvable models in statistical mechanics are those which can be studied with the Bethe ansatz approach (in its modern sense, based on the Yang-Baxter equations). A loose definition, without real intrinsic meaning, would be that "they can be expressed explicitly in terms of some special functions". For a deeper understanding, the reader is advised to read the review [48].

Many different approaches have been used: determinantal processes, orthogonal polynomials, contour integral formulas, algebraic and combinatorial methods, representation theory. The RSK correspondence is a combinatorial mapping which plays a fundamental role in the theory of Young tableaux, symmetric functions and representation theory. It has been used by Johansson to express the law of the last passage time in 2-dimensional oriented percolation with geometrically distributed passage times in terms of Schur polynomials, which also have an explicit determinantal expression. In the special case of i.i.d. geometric weights, Johansson [145] proved that the last passage time to a distant point on the diagonal, appropriately scaled, converges in law to the Tracy-Widom distribution for the GUE. Last passage percolation with exponentially distributed passage times is another solvable model.

Among solvable models we mention the semidiscrete polymer (a continuous-time random walk paths in Brownian environment) [180, 190], the log-gamma polymer [211], and more generally Macdonald processes, which are probability measures on Gelfand-Tsetlin patterns yielding a common algebraic framework [47]. Allowing zero temperature, we have the above percolation model (geometric or exponential weights), and totally asymmetric exclusion process (TASEP).

We add a few words on the semidiscrete polymer, which is indexed by $\mathbb{R}_+ \times \mathbb{N}$ instead of $\mathbb{N} \times \mathbb{N}$ as in the models considered here with $d = 1$ and after a 45° rotation. The disorder is given by (independent) Brownian motions on each line. This model is the positive temperature version of a (semi-discrete) last passage percolation problem known as O'Connell-Yor model but originating in queueing theory [135]. It can viewed as the limit of the discrete polymer where the random walk has a strong asymmetry, by rescaling the horizontal coordinate as the drift gives an overwhelming probability for the walk to use right steps rather than upwards steps. It is known [47, 220] that the semidiscrete polymer has Tracy-Widom fluctuations. Coming back to the discrete model with general weights, the authors in [19] prove that volume fluctuations of the discrete polymer has the same limit fluctuations close to the axis, i.e., that for small enough $a \in (0, 1)$ the variable $\ln Z_{n,[n^a]}$ after suitable rescaling converges to the Tracy-Widom distribution. This is similar to the zero temperature case [44].

The log-gamma polymer is a valuable source of amazing computations and fine results. It has attracted a strong interest and inventive techniques, and it is impossible to cite all references. Let me mention just a few: large deviations of the partition function [116], computations of Busemann functions [117], GUE Tracy-Widom fluctuations for Z_n at scale $n^{1/3}$ [49]: for the model without boundaries and with μ small enough, as $n \to \infty$,

$$[-\Psi_2(\mu/2)]^{1/3} \frac{\ln Z_{n,n} + 2n\Psi_0(\mu/2)}{n^{1/3}} \xrightarrow{\text{law}} F_{\text{GUE}} \qquad (7.35)$$

where F_{GUE} denotes the GUE-Tracy-Widom distribution. In [86], the authors define the geometric Robinson-Schensted-Knuth (RSK) correspondence, prove that the pushforward of a class of product inverse-gamma probability measures under the geometric RSK mapping, so-called Whittaker measure, has a nice integral representation. They deduce an explicit formula for the Laplace transform of the partition function for a finite size.

Chapter 8
Kardar-Parisi-Zhang Equation and Universality

In this section we consider $\boxed{d = 1}$ only.

Polymer models belong to Kardar-Parisi-Zhang (KPZ) universality class, which is an extended family of models (kinetically roughened surfaces) which all share some non-Gaussian scaling limits and statistics, characterized by a few exponents different from the usual ones. Some other familiar members are some interacting particle systems (exclusion processes SSEP, ASEP, TASEP, q-TASEP), stochastic growth (random deposition, ballistic aggregation, polynuclear growth, Eden and Richardson model), first/last passage percolation, and some stochastic PDE's (stochastic Burgers equation, stochastic reaction-diffusion equations, stochastic Hamilton-Jacobi equations). An elementary introduction [84] aiming at non-specialists mathematicians covers some of the above models, and can be completed by a more physics-oriented, but still pedestrian, account [155].

Another member of this class is the KPZ equation, which is a limit of polymer models. To introduce the equation, we give a short heuristic of how it relates to polymers. The aim is to approximate the P2P partition function of a discrete polymer[1]

$$P[\exp\{\beta \sum_{t=1}^{n} \omega(t, S_t)\}; S_n = y], \quad n \in \mathbb{N}, y \in \mathbb{Z},$$

by a continuous analogue. A natural candidate could be

$$\mathfrak{Z}(t, x) = E_{0,0}^{t,x}\left[\exp\{\beta \int_0^t \dot{\mathscr{W}}(s, B(s))ds\}\right] \times g(t, x), \quad t > 0, x \in \mathbb{R},$$

---

[1]Recall the classical notation $P[X; A] = P[X\mathbf{1}_A]$ for the expectation of a r.v. X on the event A.

© Springer International Publishing AG 2017
F. Comets, *Directed Polymers in Random Environments*, Lecture Notes
in Mathematics 2175, DOI 10.1007/978-3-319-50487-2_8

with $\dot{\mathcal{W}}$ a space-time noise, B a Brownian bridge[2] starting at the origin at time 0 and ending in x at time t and $g(t,x)$ the Gaussian density of $\mathcal{N}(0,t)$. Assuming that the noise is smooth, the Feynman-Kac implies that $\mathfrak{Z}$ solves

$$\frac{\partial \mathfrak{Z}}{\partial t} = \frac{1}{2}\frac{\partial^2 \mathfrak{Z}}{\partial x^2} + \beta \mathfrak{Z}\dot{\mathcal{W}} \, ,$$

with initial value $\lim_{t \searrow 0} \mathfrak{Z}(t,x)dx = \delta_0(dx)$. In fact, we are concerned with its logarithm $\mathfrak{H}(t,x) = \ln \mathfrak{Z}(t,x)$, which is easily seen to solve KPZ equation (8.1) below. All these calculations remain only formal for us since we are interested in non-smooth noise $\dot{\mathcal{W}}$.

8.1 Kardar-Parisi-Zhang Equation

The KPZ equation is a stochastic partial differential equation, parabolic and semilinear,

$$\frac{\partial \mathcal{H}}{\partial t}(t,x) = \frac{1}{2}\frac{\partial^2 \mathcal{H}}{\partial x^2}(t,x) + \frac{1}{2}\left(\frac{\partial \mathcal{H}}{\partial x}(t,x)\right)^2 + \beta \dot{\mathcal{W}}(t,x) \, , \tag{8.1}$$

with $\mathcal{W}$ a space-time gaussian white noise ($t \geq 0, x \in \mathbb{R}$). It is a phenomenological equation, introduced in 1986 by Kardar et al. in a celebrated[3] paper [151]. The solution $\mathcal{H}(t,x)$ can be interpreted as the height function at location x and time t, evolving according to three salient features:

1. diffusion and spreading of the mass according to the Laplace operator;
2. local growth increasing with the slope via a quadratic non-linearity: the larger the absolute value of the gradient, the larger the increase;
3. random forcing uncorrelated in time and space.

The smoothing in (1) will tend to make the landscape flat, the local growth in (2) makes holes disappear and cliffs go forward; in contrast the noise term in (3) adds new irregularities and results in roughening.

The noise is the distribution-valued Gaussian field on $\mathbb{R}^+ \times \mathbb{R}$ with mean 0 and covariance

$$\mathbb{E}[\dot{\mathcal{W}}(t,x)\dot{\mathcal{W}}(s,y)] = \delta(t-s)\delta(x-y).$$

---

[2] We denote by $E_{s,x}^{t,y}$ the expectation for the Brownian bridge starting from x at time s and ending at y at time $t > s$.

[3] According to Google Scholar, this paper has received 4500 citations in date of June 30, 2016. Nowadays there are many mathematical papers on KPZ, we will restrict ourselves here on the connections to polymers.

(We denote by $\mathbb{E}$ the expectation operator on the probability space where the white noise is defined.) This means that the noise acts on smooth functions $f(t, x)$, and denoting $\int f\dot{\mathcal{W}} = \int f\dot{\mathcal{W}} dtdx$ its action, the random variables $\{\int f\dot{\mathcal{W}} ; f \in L^2(\mathbb{R}^+ \times \mathbb{R})\}$ is a Gaussian family with mean 0 and

$$\mathbb{E}\left[\int f\dot{\mathcal{W}} \int g\dot{\mathcal{W}}\right] = \int_{\mathbb{R}^+ \times \mathbb{R}} f(t, x)g(t, x)dtdx. \tag{8.2}$$

The noise can be constructed as a convergent series in a negative Sobolev space [196] and the construction of stochastic integrals is given there in details: Denote by $\Omega_{\dot{\mathcal{W}}}$ the space where the noise is defined, and as usual $\mathcal{F}_t^{\dot{\mathcal{W}}}$ the σ-field generated by the collection of variables $\int f\dot{\mathcal{W}}$ with f supported in the time interval $[0, t]$. Let $\mathcal{P}$ be the σ-field on $\mathbb{R}^+ \times \mathbb{R} \times \Omega_{\dot{\mathcal{W}}}$ generated by non-anticipating simple functions

$$f(t, x, \omega) = \sum_{i=1}^{n} X_i(\omega)\mathbf{1}_{(s_{i-1}, s_i]}(t)\phi_i(x),$$

for $0 = s_0 < s_1 < \ldots$, $X_i \in L^2(\mathcal{F}_{s_{i-1}}^{\dot{\mathcal{W}}})$ and $\phi_i \in L^2(\mathbb{R})$. For compactly supported ϕ_i's, the stochastic integral is defined as

$$\int f(t, x, \omega)\dot{\mathcal{W}}(t, x)dtdx = \sum_{i=1}^{n} X_i(\omega) \int_{\mathbb{R}^+ \times \mathbb{R}} \mathbf{1}_{(s_{i-1}, s_i]}(t)\phi_i(x)\dot{\mathcal{W}}(t, x)dtdx.$$

This is an isometry, and it extends by density to $f \in L^2(\mathbb{R}^+ \times \mathbb{R} \times \Omega_{\dot{\mathcal{W}}}, \mathcal{P})$, into a linear isometry $f \mapsto \int f\dot{\mathcal{W}}$ with

$$\mathbb{E}\left[\int f\dot{\mathcal{W}}\right] = 0, \qquad \mathbb{E}\left[\left(\int f\dot{\mathcal{W}}\right)^2\right] = \int \mathbb{E}f^2(t, x)dtdx.$$

At a formal level, we see that the gradient $u(t, x) = \frac{\partial \mathcal{H}}{\partial x}(t, x)$ of the height function from (8.1) should solve, if things were nice, the stochastic Burgers equation with positive viscosity

$$\frac{\partial u}{\partial t} = \frac{1}{2}\frac{\partial^2 u}{\partial x^2} + \frac{1}{2}\frac{\partial(u^2)}{\partial x} + \beta\frac{\partial \dot{\mathcal{W}}}{\partial x}.$$

Rescaling time and space produces arbitrary coefficients in front of the three terms in the RHS of (8.1). The natural scale for time/space/fluctuations is 3:2:1, in contrast with the 4:2:1 scale in the Edwards-Wilkinson model [21, 105, 210] where the nonlinearity is suppressed (i.e., the middle term in the right-hand side is missing). Thinking to the two models as dynamic of an interface separating two phases, the difference is that Edwards-Wilkinson model is the case of two equivalent phases, though KPZ equation describes situations where one phase is favored: the middle term pushes the interface upwards. See [83, 196] for details, and [229] for a physics point-of-view.

8.2 Hopf-Cole Solution

It is well known that homogeneous products of Bernoulli are invariant (even reversible) for the Simple Exclusion Process (SEP). Using a discrete Cole-Hopf transformation due to Gärtner, Bertini and Giacomin [32] proved that renormalized WASEP (weakly asymmetric simple exclusion process) converges to stochastic Burgers equation, and then Gaussian white noise—coming as the scaling limit of the Bernoulli process—has to be invariant (and reversible) for stochastic Burgers: hence, by integration in space, they derive that Brownian motion is invariant (up to a shift) by KPZ. We will indicate another proof later in Remark 8.3, but write a formal statement already here.

Proposition 8.1 ([32, 113]; See Also [22])   *The solution $\mathcal{H}$ of* (8.1) *starting from*

$$\mathcal{H}(0, \cdot) = \sqrt{2}\beta B(\cdot)$$

with B a double-sided Brownian motion with diffusion coefficient 1, has the following stationarity property:

$$\left(\mathcal{H}(t,x) - \mathcal{H}(t,0)\right)_{x\in\mathbb{R}} \stackrel{\text{law}}{=} \left(\mathcal{H}(0,x)\right)_{x\in\mathbb{R}}.$$

Since the Brownian motion is not differentiable, the non-linearity in KPZ equation is not well defined at equilibrium. It can be checked that the same problem occurs far from equilibrium as well, and the term $(\partial_x\mathcal{H})^2$ does not make sense. Thus it is not even clear what the meaning of Eq. (8.1) is. Bertini and Giacomin proposed in [32] to interpret the equation via the Hopf-Cole transformation

$$\mathcal{H}(t,x) = \ln \mathcal{Z}(t,x) , \tag{8.3}$$

with $\mathcal{Z} = \mathcal{Z}(t,x)$ solution of the Stochastic Heat Equation (SHE),

$$\frac{\partial \mathcal{Z}}{\partial t} = \frac{1}{2}\frac{\partial^2 \mathcal{Z}}{\partial x^2} + \beta \mathcal{Z}\dot{\mathcal{W}} . \tag{8.4}$$

(This equation is linear, it has a multiplicative noise.) They are many reasons to support this choice.

1. If $\dot{\mathcal{W}}$ was smooth, then this would be just the standard Hopf-Cole transformation, making (8.4) and (8.1) equivalent.

2. If we smooth out the noise with a small regularization ε, SHE has a solution which logarithm $\mathcal{H}_\varepsilon$ solves an equation

$$\frac{\partial}{\partial t}\mathcal{H}_\varepsilon(t,x) = \frac{1}{2}\frac{\partial^2}{\partial x^2}\mathcal{H}_\varepsilon(t,x) + \left(\frac{1}{2}\left(\frac{\partial}{\partial x}\mathcal{H}_\varepsilon(t,x)\right)^2 - C_\varepsilon\right) + \beta\dot{W}_\varepsilon(t,x),\quad (8.5)$$

with a diverging $C_\varepsilon \to +\infty$ (cf. [113, 196] for details).
3. It yields the correct exponents $\chi^\perp, \chi^\parallel$.

Equation (8.4) being linear, the solution is formally given by the Feynman-Kac formula, but one runs into serious problem all along the route, see (8.11) below.

Stochastic heat equation (8.4) can be rewritten in Duhamel form[4] as

$$\mathcal{Z}(t,x) = \int_{\mathbb{R}} g(t,x-y)\mathcal{Z}(0,y)dy + \beta\int_0^t\int_{\mathbb{R}} g(t-s,x-y)\mathcal{Z}(s,y)\dot{W}(s,y)dsdy,$$
$$(8.6)$$

using the heat kernel $g(t,x) = (2\pi t)^{-1/2}\exp\{-x^2/2t\}$.

Definition 8.1 A mild solution of SHE (8.4) is a progressively measurable process $\mathcal{Z}(t,x)$ such that

$$\int_0^t\int_{\mathbb{R}} g(t-s,x-y)^2\mathbb{E}[\mathcal{Z}^2(s,y)]dsdy < \infty, \quad (8.7)$$

and such that (8.6) holds.

It is not difficult to show that, for an $\mathscr{F}_0^W$-measurable initial condition $z_0 = \mathcal{Z}(0,\cdot)$ with

$$\mathbb{E}\sup_{A\subset[-n,n]}\int_A z_0(x)dx \leq Ce^{C'n}, \quad (8.8)$$

there exists a unique strong solution of (8.4). (Note that z_0 can be a measure.) Iterating once again in (8.6) we obtain

$$\mathcal{Z}(t,x) = \int_{\mathbb{R}} g(t,x-y_0)z_0(y_0)dy_0$$
$$+ \beta\int_0^t\int_{\mathbb{R}^2} g(t-s_1,x-y_1)g(s_1,y_1-y_0)z_0(y_0)\dot{W}(s_1,y_1)ds_1dy_0dy_1$$
$$+ \beta^2\int_{0\leq s_1<s_2\leq t}\int_{\mathbb{R}^2} g(t-s_2,x-y_2)g(s_2-s_1,y_2-y_1)\mathcal{Z}(s_1,y_1)$$
$$\times \dot{W}(s_1,y_1)\dot{W}(s_2,y_2)ds_1ds_2dy_1dy_2.$$

---

[4]See (A.20) in Appendix, taking $L = \frac{1}{2}\frac{\partial^2}{\partial x^2}$ so that $e^{tL}(x,y) = g(t,y-x)$, and $F(t) = \beta\dot{W}\mathcal{Z}(t,\cdot)$.

Iterating indefinitely we see that the mild solution is given by the **chaos representation**

$$\mathscr{Z}(t,x) = g_t * z_0 + \sum_{k=1}^{\infty} \beta^k \int_{\Delta_{k,t}} \int_{\mathbb{R}^{k+1}} \mathbf{g}_k(\mathbf{s}, y_0, \mathbf{y}, x) z_0(y_0) \dot{\mathscr{W}}^{\otimes k}(\mathbf{s}, \mathbf{y}) dy_0 d\mathbf{y} d\mathbf{s} ,$$

$$(8.9)$$

where

$$\Delta_{k,t} = \{0 < s_1 < s_2 \ldots < s_k < t\},$$

and for $\mathbf{s} = (s_1, \ldots, s_k)$, $\mathbf{y} = (y_1, \ldots y_k)$,

$$\mathbf{g}_k(\mathbf{s}, y_0, \mathbf{y}, x) = g(t - s_k, x - y_k) g(s_k - s_{k-1}, y_k - y_{k-1}) \ldots g(s_1, y_1 - y_0).$$

(Observe that the general term in the above series is equal to $g_t * z_0$ when $k = 0$, so the right-hand side of (8.9) can be written as a sum over $k \geq 0$.) The multiple integral in (8.9) has been obtained as an iterated stochastic integral, however we write it as a product integral on $[0, t]^k \times \mathbb{R}^k$, so a proper definition of the latter is needed. The construction is similar to the $k = 1$ case, except that some care is necessary along the diagonal and that one starts with integrating symmetric function. The reader is referred to [143] for a complete account. In particular, for any $\mathbf{f} \in L^2(\Delta_{k,t} \times \mathbb{R}^k)$, the stochastic integral $\int_{\Delta_{k,t}} \int_{\mathbb{R}^k} \mathbf{f}(\mathbf{s}, \mathbf{y}) \dot{\mathscr{W}}^{\otimes k}(\mathbf{s}, \mathbf{y}) d\mathbf{y} d\mathbf{s}$ is a mean zero random variable with variance $\|\mathbf{f}\|_{L^2}^2$, and the covariance structure is

$$\mathbb{E}\left[\int_{\Delta_{k,t}} \int_{\mathbb{R}^k} \mathbf{f}_1 \dot{\mathscr{W}}^{\otimes k} \times \int_{\Delta_{k,t}} \int_{\mathbb{R}^\ell} \mathbf{f}_2 \dot{\mathscr{W}}^{\otimes k} \right] = \begin{cases} \langle \mathbf{f}_1, \mathbf{f}_2 \rangle_{L^2(\Delta_{k,t} \times \mathbb{R}^k)} & \text{if } k = \ell, \\ 0 & \text{if } k \neq \ell. \end{cases} \quad (8.10)$$

The series (8.9) is a converging series in $L^2(\mathbb{P})$—in fact it is an orthogonal series—and the sum indeed defines a mild solution of SHE. (cf. [196].) It has been proved [176, 182] that the solution of (8.4) starting from $z_0 \geq 0$ and $\int_R z_0 > 0$ satisfies $\mathscr{Z}(t,x) > 0$ $\forall x$ with probability one, showing it is legitimate to consider $\mathscr{H} = \ln \mathscr{Z}$.

Proposition 8.2 *The stochastic heat equation (8.4) has a unique mild solution given by (8.9) for all initial datum satisfying (8.8).*

By this route we only give a weak definition of what a solution to SHE equation is, but this is enough for polymers. Giving a rigorous meaning to KPZ equation and further understanding the fine and strong properties is more of a challenge, and it has culminated recently with Martin Hairer's paper [128] and the theory of regularity structures [130], introducing a new concept which essentially provides a pathwise approach, together with a very detailed approximation theory. We also mention alternative viewpoints of paracontrolled distributions [125] and energy solutions [122]. For a first overview we refer to [89, 129].

8.3 The Continuum Random Polymer (CRP)

We would like to define a polymer model on continuous paths $\mathbf{x} = (x(t) : t \in [0, T])$ taking values in $\mathbb{R}$ with density proportional to

$$\exp \left\{ \beta \int_0^T \dot{\mathcal{W}}(t, x(t)) dt \right\}$$

w.r.t. the Brownian motion, or on paths $\mathbf{x}$ constrained to $x(T) = x$ with this density w.r.t. the Brownian bridge. This seems doubtful, since the integral along the Brownian path is not well defined (with only a one-dimensional integral). In fact, this is not possible at all, since, as we see below, the model we look for is not absolutely continuous to Wiener measure. [We can already guess it from the divergence of the normalization C_ε in (8.5).] However we note that the chaos expansion (8.9) with $z_0 = \delta_0$ can be written as an expectation over Brownian bridges $x(t)$,

$$\mathcal{L}(T, x) = E_{0,0}^{T,x} \left[: \exp : \left\{ \beta \int_0^T \dot{\mathcal{W}}(t, x(t)) dt \right\} \right] \times g(T, x) \tag{8.11}$$

with $:$ exp $:$ the Wick exponential [143]. Compared to the usual exponential $\exp(u) = \sum_{k \geq 0} u^k / k!$, the Wick exponential of the stochastic integral $:$ exp $:$ $\left\{ \int_0^T \dot{\mathcal{W}}(t, x(t)) dt \right\}$ is given by the same series where powers are replaced by Wick products,

$$: \left(\int_0^T \dot{\mathcal{W}}(t, x(t)) dt \right)^k : \; = k! \times \int_{\Delta_{k,T}} \prod_{j=1}^k \dot{\mathcal{W}}(t_j, x(t_j)) dt_j \ ,$$

taking a special care of the diagonal terms in the k-fold integral. Integrating the sum w.r.t. the Brownian bridge, one recovers the RHS of (8.9). This shows (8.11).

Coming back to the task of defining the continuum random polymer, a way to get around the above issue is to take advantage of the Markov structure by considering $\mathcal{L}(s, x; t, y)$ the solution of the SHE starting from location x at time s, i.e., the unique solution of

$$\begin{cases} \dfrac{\partial \mathcal{L}}{\partial t} = \dfrac{1}{2} \dfrac{\partial^2 \mathcal{L}}{\partial x^2} + \beta \mathcal{L} \dot{\mathcal{W}} & (t > s, y \in \mathbb{R}), \\ \lim \mathcal{L}(s, x; t, y) dy = \delta_x(dy) & \text{as } t \searrow s. \end{cases} \tag{8.12}$$

These (unnormalized) transition kernels satisfy Chapman-Kolmogorov equation (cf. Property 7 below), allowing to define the continuum random polymer as a Markov process.

8.3.1 Properties of $\mathscr{L}_\beta$

We first summarize the necessary properties of the kernels $\mathscr{L}(s,x;t,y) = \mathscr{L}_\beta(s,x;t,y)$ for which we recall the dependence in β only when necessary.

Proposition 8.3 ([5]) *The solution $\mathscr{L}_\beta(s,x;t,y)$ of (8.12) has the following properties:*

1. *$\mathscr{L}_\beta(s,x;t,y)$ is given by $\mathscr{L}(t-s,y-x)$, where $\mathscr{L}$ has the form of (8.9) with a time-space shift by the vector (s,x), namely*

$$\mathscr{L}_\beta(s,x;t,y) = \sum_{k=0}^{\infty} \beta^k \int_{\Delta_{k,s,t}} \int_{\mathbb{R}^k} \prod_{\ell=1}^{k+1} g(s_\ell - s_{\ell-1}, y_\ell - y_{\ell-1}) \dot{W}(s_\ell, y_\ell)\, ds_1 dy_1 \ldots ds_k dy_k$$

with $\Delta_{k,s,t} = \{(s_\ell)_{\ell=1}^{k} : s < s_1 < \ldots < s_k < t\}$, $s_0 = s$, $y_0 = x$, $s_{k+1} = t$ and $y_{k+1} = y$.

2. *$\mathscr{L}_\beta(s,x;t,y)$ is a.s. continuous in its four variables; $\mathscr{L}_\beta(s,x;t,y) > 0$ a.s.*
3. *$\mathbb{E}\mathscr{L}_\beta(s,x;t,y) = g(t-s, y-x)$;*
4. *Stationarity: $\mathscr{L}_\beta(s+u, x+v; t+u, y+v) \overset{\text{law}}{=} \mathscr{L}_\beta(s,x;t,y)$;*
5. *Scaling: $\mathscr{L}_\beta(r^2 s, rx; r^2 t, ry) \overset{\text{law}}{=} r^{-1}\mathscr{L}_{\sqrt{r}\beta}(s,x;t,y)$ for $r > 0$;*
6. *The law of $\mathscr{L}_\beta(s,x;t,y)/g(t-s, y-x)$ does not depend on x, y.*
7. *Chapman-Kolmogorov: for $s \leq \tau \leq t$,*

$$\int_{\mathbb{R}} \mathscr{L}_\beta(s,x;\tau,\xi)\mathscr{L}_\beta(\tau,\xi;t,y)d\xi = \mathscr{L}_\beta(s,x;t,y). \tag{8.13}$$

Proof (Sketch of Proof) Claim 1 was already discussed. Regularity in Claim 2 is standard for a linear SPDE [237], positivity was shown in [176, 182]. Claim 3 is clear from the orthogonal sum (8.9). Property 4 follows from translation invariance of the white noise. We now prove Claim 5 for $s = 0, x = 0$, which is not a restriction in view of the above properties. We use the Brownian scaling and the white noise scaling in representation (8.11): Under $E_{0,0}^{r^2 t, rx}$, the process $B^{(r)}$ defined by $B^{(r)}(s) = r^{-1}B(r^2 s)$ is a Brownian bridge form $(0,0)$ to (t,x) by Brownian scaling; Moreover, fixing $a, b > 0$,

$$\dot{W}^{(a,b)}(t,x) = \dot{W}(at, bx) \quad \text{defines a process} \quad \dot{W}^{(a,b)} \overset{\text{law}}{=} (ab)^{-1/2}\dot{W}. \tag{8.14}$$

Indeed, for a test function f, by definition

$$\int f(t,x)\dot{W}^{(a,b)}(t,x)dtdx = (ab)^{-1}\int f(s/a, y/b)\dot{W}(s,y)dsdy$$

is Gaussian with variance

$$\mathbb{E}\left(\int f\dot{\mathscr{W}}^{(a,b)}\right)^2 = (ab)^{-2}\int f(s/a, y/b)^2 ds dy = (ab)^{-1}\int f^2.$$

Thus,

$$\frac{\mathscr{Z}_\beta(0,0;r^2t, rx)}{g(r^2t, rx)} = E_{0,0}^{r^2t, rx}\left[:\exp:\{\beta\int_0^{r^2t}\dot{\mathscr{W}}(u, B(u))du\}\right]$$

$$= E_{0,0}^{r^2t, rx}\left[:\exp:\{\beta\int_0^{r^2t}\dot{\mathscr{W}}(u, rB^{(r)}(u/r^2))du\}\right]$$

$$\overset{\text{Brownian scal.}}{=} E_{0,0}^{t,x}\left[:\exp:\{\beta r^2\int_0^t\dot{\mathscr{W}}(r^2s, rB(s))ds\}\right]$$

$$= E_{0,0}^{t,x}\left[:\exp:\{\beta r^2\int_0^t\dot{\mathscr{W}}^{(r^2, r)}(s, B(s))ds\}\right]$$

$$\overset{(8.14)}{=} E_{0,0}^{t,x}\left[:\exp:\{\beta r^{1/2}\int_0^t\dot{\mathscr{W}}(s, B(s))ds\}\right] \quad \text{in law.}$$

Since $g(r^2t, rx) = r^{-1}g(t,x)$, Claim 5 follows.

We now prove Claim 6 for $s = 0, x = 0$ without loss of generality. The Brownian bridge from $(0,0)$ to (t,y) being a shift of the bridge from $(0,0)$ to $(t,0)$ by a straight line with slope $a = y/t$, we can express the ratio under consideration with the shifted environment

$$\dot{\mathscr{W}}_a(s,x) = \dot{\mathscr{W}}(s, x + as),$$

which is still a Gaussian white noise, with mean 0 and covariance as in (8.2): by (8.11),

$$\frac{\mathscr{Z}_\beta(0,0;t,y)}{g(t,y)} = E_{0,0}^{t,y}\left[:\exp:\{\beta\int_0^t\dot{\mathscr{W}}(s, x(s))ds\}\right]$$

$$= E_{0,0}^{t,0}\left[:\exp:\{\beta\int_0^t\dot{\mathscr{W}}(s, x(s) + as)ds\}\right]$$

$$= E_{0,0}^{t,0}\left[:\exp:\{\beta\int_0^t\dot{\mathscr{W}}_a(s, x(s))ds\}\right]$$

$$\overset{\text{law}}{=} \frac{\mathscr{Z}_\beta(0,0;t,0)}{g(t,0)},$$

since $\dot{\mathscr{W}}_a \overset{\text{law}}{=} \dot{\mathscr{W}}$. This proves Property 6.

Finally, stochastic heat equation (8.4) being linear, the LHS of (8.13) solves SHE on $[s, \infty)$ starting from δ_x, and by uniqueness of the solution, it has to be equal to the RHS. (Actually, one needs to be more careful: the LHS satisfies (8.6) at all rational points, and continuity in Property 2 implies it is a mild solution.) The reader is referred to [5] for the numerous necessary details. $\qquad\square$

Remark 8.1 From the above proof we see how useful and convenient the representation (8.11) is.

For further purpose, introduce the notation

$$\mathscr{L}(s, x; T, *) = \int_{\mathbb{R}} \mathscr{L}(s, x; T, y)dy , \qquad (8.15)$$

and the process

$$A_\beta(t, y) \overset{\text{def.}}{=} \ln \frac{\mathscr{L}_\beta(0, 0; t, y)}{g(t, y)} = \ln E_{0,0}^{t,y}\left[: \exp : \left\{ \beta \int_0^t \dot{\mathscr{W}}(s, x(s))ds \right\} \right]. \qquad (8.16)$$

It has two immediate, though remarkable, properties.

Proposition 8.4

(i) *Scaling: the processes $(A_\beta(t, x))_{t,x}$ and $(A_1(\beta^4 t, \beta^2 x))_{t,x}$ have the same law.*
(ii) *Space-shift invariance: for all β, t, the process $A_\beta(t, \cdot)$ is stationary in space.*

Proof

(i) follows from the scaling relation 5 in Proposition 8.3.

$$A_\beta(t, x) = \ln \frac{\mathscr{L}_\beta(0, 0; t, x)}{g(t, x)}$$

$$\overset{\text{law}}{=} \ln \frac{\beta^2 \mathscr{L}_1(0, 0; \beta^4 t, \beta^2 x)}{\beta^2 g(\beta^4 t, \beta^2 x)}$$

$$= A_1(\beta^4 t, \beta^2 x) .$$

(ii) From the proof of Property 7 in Proposition 8.3, we see that for fixed t and $z = bt$, the processes

$$\left(\frac{\mathscr{L}_\beta(0, 0; t, y)}{g(t, y)} \right)_{y \in \mathbb{R}} = F_t(\dot{\mathscr{W}}) , \qquad \left(\frac{\mathscr{L}_\beta(0, 0; t, z + y)}{g(t, z + y)} \right)_{y \in \mathbb{R}} = F_t(\dot{\mathscr{W}}_b) ,$$

are the same functional taken on fields $\dot{\mathscr{W}}$ and $\dot{\mathscr{W}}_b$ sharing the same distribution. Hence, they have themselves the same law, and for fixed t the process $A_\beta(t, \cdot)$ is *stationary in space*. $\qquad\square$

On the contrary to the customary use to set the parameter β equal to 1, we keep it explicit since it has a central meaning for polymers.

8.3.2 The Law of the Continuum Random Polymer

By Chapman-Kolmogorov, for almost every environment, the family of distributions

$$P_{0,0}^{T,\beta}(x(s_1) \in dx_1, \ldots, x(s_k) \in dx_k) = \frac{\prod_{i=1}^{k+1} \mathscr{Z}_b(s_{i-1}, x_{i-1}; s_i, x_i)}{\mathscr{Z}_b(0, 0; T, *)} dx_1 \ldots dx_k,$$

(8.17)

for $0 < s_1 < \ldots < s_k < T$ and with $s_0 = 0, x_0 = 0, s_{k+1} = T, x_{k+1} = *$, is a consistent family. By Kolmogorov extension, it defines a Markov process on $[0, T]$.

Definition 8.2 The point-to-line KPZ continuum random polymer with time horizon T is the Markov process (for a fixed environment $\mathscr{W}$) on paths $[0, T] \to \mathbb{R}$ with finite dimensional marginal distributions given by (8.17). The point-to-point polymer is defined analogously.

We denote it by $\mathscr{X}_{0,0}^T$ (and $\mathscr{X}_{0,0}^{T,x}$ the bridge version), it depends on the environment $\dot{\mathscr{W}}$, and its restriction to a smaller time interval $[0, S]$ is *not* equal to $\mathscr{X}_{0,0}^S$ in distribution. We now mention the main properties of the KPZ continuum random polymer.

Proposition 8.5 ([5, Theorem 4.3])

1. $P_{0,0}^{T,\beta}$ *is supported by* $\mathscr{C}[0, T]$, *and even by* $(1/2)^-$ *Holder-continuous functions;*
2. *The quadratic variation of* $\mathscr{X}_{0,0}^T$ *and* $\mathscr{X}_{0,0}^{T,x}$ *is equal to* T;
3. $P_{0,0}^T$ *is a.s. singular to the Wiener measure (and* $P_{0,0}^{T,x}$ *is singular to the bridge measure).*

Moreover, it is proved there that $\mathscr{X}_{0,0}^T$ (and $\mathscr{X}_{0,0}^{T,x}$ similarly) is a diffusion process in random environment. Letting

$$\mathscr{H}(s, x; T, *) = \ln \mathscr{Z}_\beta(s, x; T, *),$$

we can see that $P_{0,0}^T$ is *formally* the law of the solution $\mathscr{X}$ on $[0, T]$ of the stochastic differential equation

$$d\mathscr{X}(s) = \frac{\partial}{\partial x}\mathscr{H}(s, \mathscr{X}(s); T, *)ds + dB(s), \qquad \mathscr{X}(0) = 0, \qquad (8.18)$$

with B a Brownian motion. Observe that this equation is only formal since the space derivative does not exist. The derivation can be sketched as follows. For $0 \le s < t \le T$ denote by $Q^T_{s,t}$ the transition operator

$$Q^T_{s,t} f(x) \stackrel{\text{def.}}{=} P^T_{0,0}\left[f(\mathscr{X}(t)) \,\big|\, \mathscr{X}(s) = x\right]$$

$$= \mathscr{L}(s,x;T,*)^{-1} \int f(y) \mathscr{L}(s,x;t,y) \mathscr{L}(t,y;T,*) dy$$

$$= \mathscr{L}(s,x;T,*)^{-1} E\left[f(B(t)) \frac{\mathscr{L}(s,x;t,B(t))}{g(t-s,B(t)-x)} \mathscr{L}(t,B(t);T,*) \,\big|\, B(s) = x\right]$$

where E is the expectation w.r.t. the Brownian motion B. We can write

$$Q^T_{s,t} f(x) - f(x) = \mathscr{L}(s,x;T,*)^{-1}$$

$$\times E\left[[f(B(t)) - f(x)] \frac{\mathscr{L}(s,x;t,B(t))}{g(t-s,B(t)-x)} \mathscr{L}(t,B(t);T,*) \,\big|\, B(s) = x\right],$$

where the ratio $\mathscr{L}(s,x;t,B(t))/g(t-s,B(t)-x)$ converges to 1 as $t \searrow s$, thus

$$\lim_{t \searrow s} \frac{Q^T_{s,t} f(x) - f(x)}{t-s}$$

$$= \mathscr{L}(s,x;T,*)^{-1} \lim_{t \searrow s}(t-s)^{-1} E\left[[f(B(t)) - f(x)]\mathscr{L}(t,B(t);T,*) \,\big|\, B(s) = x\right]$$

$$= \mathscr{L}(s,x;T,*)^{-1} \lim_{t \searrow s} \frac{E\left[[f(B(t))\mathscr{L}(t,B(t);T,*) \,\big|\, B(s) = x] - f(x)\mathscr{L}(s,x;T,*)\right]}{t-s}$$

$$= \mathscr{L}(s,x;T,*)^{-1} \left(\frac{1}{2}\frac{\partial^2}{\partial x^2}[f(x)\mathscr{L}(s,x;T,*)] + f(x)\frac{\partial \mathscr{L}(s,x;T,*)}{\partial s}\right).$$

The last equality would follow from Ito's formula if the functions were smooth; we recall this is not the case for $\mathscr{L}$, and our discussion remains formal! Now, again if things were nice, we would use the Kolmogorov backward equation satisfied by the function $\mathscr{L}(s,x;T,*)$ (as well as $\mathscr{L}(s,x;T,y)$):

$$\frac{1}{2}\frac{\partial^2 \mathscr{L}(s,x;T,*)}{\partial x^2} + \frac{\partial \mathscr{L}(s,x;T,*)}{\partial s} = 0.$$

Therefore, the infinitesimal generator $\mathscr{L}^T_s$ of the Markov process $P^T_{0,0}$ is given at time $s \in (0,T)$ by

$$\mathscr{L}^T_s f(x) = \lim_{t \searrow s} \frac{Q^T_{s,t} f(x) - f(x)}{t-s} = \frac{1}{2}f''(x) + \mathscr{L}(s,x;T,*)^{-1}\frac{\partial \mathscr{L}(s,x;T,*)}{\partial x}f'(x),$$

which is associated to the diffusion (8.18).

Coming back to mathematical grounds, we mention the construction [90] of the solution of (8.18) on the torus, in the general framework of stochastic differential equations with distributional drift. Finally, taking the asymptotic of an exact Fredholm determinant formula for the Laplace transform of the partition function, the authors in [50] show that for large time, the law of the rescaled free energy fluctuations converge to the GUE Tracy-Widom distribution.

8.4 Intermediate Disorder Regime for Lattice Polymers

We come back to the discrete polymer, at a vanishing temperature inverse β_n: recalling that the critical value is $\beta_c = 0$ in dimension $d = 1$, we hope by a fine tuning to capture the transition between weak and strong disorder. Let

$$\beta_n = n^{-1/4} \frac{1}{\sqrt{2\mathrm{Var}(\omega)}} \beta, \tag{8.19}$$

for some fixed $\beta > 0$. (The factor $1/\sqrt{2\mathrm{Var}(\omega)}$ is for simpler factors below.) We use the short notation $Z_{n,x}(\omega, \beta)$ for the point-to-point partition function

$$Z_{n,x}(\omega, \beta) = P\left[e^{\{\beta \sum_{1 \leq j \leq n} \omega(j, S_j)\}} \mathbf{1}_{S_n = x} \right].$$

Theorem 8.1 ([6]) *The normalized partition function of the discrete random polymer at intermediate disorder converges to the partition function of the continuum random polymer:*

$$Z_n(\omega, \beta_n) \exp\{-n\lambda(\beta_n)\} \xrightarrow{\mathrm{law}} \mathcal{Z}_\beta(1, *), \tag{8.20}$$

where $\mathcal{Z}_\beta(1, *) = \int \mathcal{Z}_\beta(1, x)dx$, i.e.,

$$\mathcal{Z}(t, *) = 1 + \sum_{k=1}^{\infty} \beta^k \int_{\Delta_{k,t}} \int_{\mathbb{R}^k} \mathbf{g}_k(\mathbf{s}, \mathbf{y}, *) \dot{\mathcal{W}}^{\otimes k}(\mathbf{s}, \mathbf{y}) d\mathbf{y} d\mathbf{s}, \tag{8.21}$$

where $\mathbf{g}_k(\mathbf{s}, \mathbf{y}, *) = g(s_k - s_{k-1}, y_k - y_{k-1}) \ldots g(s_1, y_1 - y_0)$ and $y_0 = 0$. Moreover, for the point-to-point partition function,

$$\frac{\sqrt{n}}{2} Z_{n, \sqrt{n}x}(\omega, \beta_n) \exp\{-n\lambda(\beta_n)\} \xrightarrow{\mathrm{law}} \frac{1}{\sqrt{2\pi}} \exp\{A_\beta(1, x) - x^2/2\} \tag{8.22}$$

where the topology is the supremum norm on bounded functions, and A_β is defined in (8.16).

The theorem illustrates the intermediate regime between weak and strong disorder since

$$\lim_{n\to\infty} Z_n(\omega, \beta_n) \exp\{-n\lambda(\beta_n)\} \quad \begin{cases} \longrightarrow & 1 \quad (\beta \searrow 0), \\ \in & (0, \infty), \ \forall \beta \in (0, \infty), \\ \longrightarrow & 0 \quad (\beta \nearrow \infty). \end{cases}$$

Proof Main steps: Without loss of generality we can assume $t = 1$, $\mathbb{P}\omega(i, x) = 0$, and $\mathrm{Var}(\omega(i, x)) = 1$.

- Let us write $e^{\beta_n \omega(i,x) - \lambda(\beta_n)} = 1 + n^{-1/4}(\beta/\sqrt{2})\zeta_n(i, x)$, where we normalize

$$\zeta_n(i, x) = n^{1/4} \frac{\sqrt{2}}{\beta} \left(e^{\beta_n \omega(i,x) - \lambda(\beta_n)} - 1\right)$$

$$= \omega(i, x) + \frac{\beta}{2\sqrt{2}n^{1/4}}[\omega(i, x)^2 - 1] + \mathcal{O}_{L^p}(n^{-1/2})$$

to have mean 0 and variance close to 1. We expand

$$Z_n(\omega, \beta_n)e^{-n\lambda(\beta_n)} = P\prod_{i=1}^n \left(1 + n^{-1/4}\frac{\beta}{\sqrt{2}}\zeta_n(i, S_i)\right) \tag{8.23}$$

$$= 1 + \sum_{k=1}^n n^{-k/4}\left(\frac{\beta}{\sqrt{2}}\right)^k \sum_{\mathbf{i} \in D_k} P\left[\prod_{j=1}^k \zeta_n(\mathbf{i}_j, S_{\mathbf{i}_j})\right]$$

with D_k the discrete simplex

$$D_k = \{\mathbf{i} = (i_1, \ldots, i_k) \in \mathbb{N}^k : 1 \le i_1 \le \ldots < i_k\}.$$

Using the transition probability kernel $p(i, x) = P(S_i = x)$ of the walk,

$$Z_n(\omega, \beta_n)e^{-n\lambda(\beta_n)} = 1 + \sum_{k=1}^n \left(\frac{\beta}{\sqrt{2}}\right)^k \sum_{\mathbf{i} \in D_k} \sum_{\mathbf{x} \in \mathbb{Z}^k} \prod_{j=1}^k \zeta_n(\mathbf{i}_j, \mathbf{x}_j)n^{-1/4}p(\Delta_j\mathbf{i}, \Delta_j\mathbf{x}),$$
$$\tag{8.24}$$

with $\mathbf{i}_0 = 0$, $\mathbf{x}_0 = 0$ and $\Delta_j\mathbf{x} = \mathbf{x}_j - \mathbf{x}_{j-1}$. Local limit theorem writes

$$p(n, x) = \frac{2}{\sqrt{n}}g(1, xn^{-1/2}) + \mathcal{O}(n^{-3/2}) \tag{8.25}$$

uniformly for x and n of the same parity.

- The correct scaling is $\beta_n = \mathcal{O}(n^{-1/4})$: Indeed, the first term ($k = 1$) in (8.24) converges to a non trivial limit, as we can check from

$$n^{-1/4} \sum_{i=1}^{n} \sum_{x\in\mathbb{Z}} \zeta_n(i,x) P(S_i = x) \xrightarrow{\text{law}} \mathcal{N}(0,\sigma^2),$$

$$\sigma^2 = \lim_n n^{-1/2} \sum_i \sum_x P(S_i = x)^2 = 2/\sqrt{\pi},$$

by computing the characteristic function and using (8.25) together with

$$g(t,x)^2 = g(t,\sqrt{2}x) \times \frac{1}{\sqrt{2\pi t}}. \tag{8.26}$$

We conclude that

$$\frac{1}{\sqrt{2}} \sum_{i=1}^{n} \sum_{x\in\mathbb{Z}} \zeta_n(i,x) n^{-1/4} P(S_i = x) \xrightarrow{\text{law}} \int_0^1 \int_{\mathbb{R}} g\dot{\mathcal{W}}. \tag{8.27}$$

- Implementing the scaling: For $\mathbf{i} \in D_k$, $(\mathbf{s},\mathbf{y}) \in \Delta_{k,t}$, set $|\mathbf{i}| = |\mathbf{s}| = |\mathbf{y}| = k$. Cells at microscopic level are $(i,x) + [0,1) \times [-1,1) \subset \mathbb{R}^+ \times \mathbb{R}$, and we rescale them diffusively at the macroscopic level:

$$R_{i,x}^n = [\frac{i}{n}, \frac{i+1}{n}) \times [\frac{x-1}{\sqrt{n}}, \frac{x+1}{\sqrt{n}})$$

and $R_{\mathbf{i},\mathbf{x}}^n = \prod_{j=1}^{|\mathbf{i}|} R_{i_j,x_j}^n$. Introduce two piecewise constant functions built from transitions and medium respectively,

$$\psi_n(\mathbf{s},\mathbf{y};p) = \sum_{(\mathbf{i},\mathbf{x})\in D_k\times\mathbb{Z}^k} \mathbf{1}_{\{(\mathbf{s},\mathbf{y})\in R_{\mathbf{i},\mathbf{x}}^n\}} n^{k/2} \prod_{j=1}^{k} p(\Delta_j\mathbf{i}, \Delta_j\mathbf{x}),$$

$$\omega_n(t,y) = \frac{n^{3/4}}{2} \sum_{(i,x)\in\mathbb{N}\times\mathbb{Z}, i\leftrightarrow x} \mathbf{1}_{\{(t,y)\in R_{i,x}^n\}} \zeta_n(i,x)$$

where the indicator functions make at most one term non zero in the sum. Then, we have the identity

$$\int_{\Delta_{k,1}} \int_{\mathbb{R}^k} \psi_n(\mathbf{s},\mathbf{y};p) \omega_n^{\otimes k}(\mathbf{s},\mathbf{y}) ds dx = \sum_{\mathbf{i}\in D_k} \sum_{\mathbf{x}\in\mathbb{Z}^k} \prod_{j=1}^{k} \zeta_n(i_j,x_j) n^{-1/4} p(\Delta_j\mathbf{i}, \Delta_j\mathbf{x})$$

$$\tag{8.28}$$

(note that the diagonal terms in the LHS are killed by the transition kernel p), and from (8.24), $Z_n(\omega, \beta_n)e^{-n\lambda(\beta_n)}$ is the sum of all these terms.
- We give a name to the right-hand side. Fix k. For $f : \mathbb{N} \times \mathbb{Z} \to \mathbb{R}$, denote

$$I_{n,k}(f) = \sum_{\mathbf{i} \in D_k} \sum_{\mathbf{x} \in \mathbb{Z}^k} \prod_{j=1}^{k} \zeta_n(\mathbf{i}_j, \mathbf{x}_j) n^{-1/4} f(\Delta_j \mathbf{i}, \Delta_j \mathbf{x}),$$

and $g_n(i, x) = \frac{2}{\sqrt{n}} g(in^{-1}, xn^{-1/2})$. Since the above sum is orthogonal, we get

$$\mathbb{E}\left[I_{n,k}(p) - I_{n,k}(g_n)\right]^2 \leq \|\psi_n(\cdot; p) - \psi_n(\cdot; g_n)\|^2_{L^2(\Delta_{k,1} \times \mathbb{R}^k)} \tag{8.29}$$

By (8.25), this difference vanishes as $n \to \infty$. For $h \in L^2([0, 1] \times \mathbb{R})$ we define a lattice function by taking the average of the function on the microscopic cell,

$$\bar{h}_n(i, x) = \frac{n^{3/2}}{2} \int_{R^n_{i,x}} h(t, z)dtdz, \qquad i \leq n, x \in \mathbb{Z}.$$

By Theorem 11.16 in [143], we have

$$2^{-k/2} \int_{\Delta_{k,1}} \int_{\mathbb{R}^k} \psi_n(\mathbf{s}, \mathbf{y}; \bar{h}_n) \omega_n^{\otimes k}(\mathbf{s}, \mathbf{y}) d\mathbf{s} d\mathbf{x} \overset{\text{law}}{\longrightarrow} \int_{\Delta_{k,1}} \int_{\mathbb{R}^k} h \dot{\mathscr{W}}^{\otimes k}.$$

Combining $\|\psi_n(\cdot; g_n) - \psi_n(\cdot; \bar{g}_n)\|_{L^2(\Delta_{k,1} \times \mathbb{R}^k)} \to 0$ as $n \to \infty$, (8.28) and (8.29), we obtain, with $\mathbf{g}_k$ given below (8.21),

$$I_{n,k}(p) \overset{\text{law}}{\longrightarrow} \int_{\Delta_{k,1}} \int_{\mathbb{R}^k} \mathbf{g}_k \dot{\mathscr{W}}^{\otimes k}.$$

Moreover, joint convergence holds: for all $K \geq 1$,

$$(I_{n,k}(f))_{k \leq K} \overset{\text{law}}{\longrightarrow} \left(\int_{\Delta_{k,1}} \int_{\mathbb{R}^k} \mathbf{g}_k \dot{\mathscr{W}}^{\otimes k} \right)_{k \leq K}.$$

For more details, see the proof of Theorem 4.3 in [6]. Thus,

$$1 + \sum_{k=1}^{K}(\beta/\sqrt{2})^k \sum_{\mathbf{i} \in D_k} \sum_{\mathbf{x} \in \mathbb{Z}^k} \prod_{j=1}^{k} \zeta_n(\mathbf{i}_j, \mathbf{x}_j) n^{-1/4} p(\Delta_j \mathbf{i}, \Delta_j \mathbf{x}) \tag{8.30}$$

$$\overset{\text{law}}{\longrightarrow} 1 + \sum_{k=1}^{K} \beta^k \int_{\Delta_{k,1}} \int_{\mathbb{R}^k} \mathbf{g}_k(\mathbf{s}, \mathbf{y}, *) \dot{\mathscr{W}}^{\otimes k}(\mathbf{s}, \mathbf{y})$$

- Commutative diagram: since

$$\sup_n \sum_{k \geq K} \sum_{\mathbf{i} \in D_k} \sum_{\mathbf{x} \in \mathbb{Z}^k} \prod_{j=1}^{k} \zeta_n(\mathbf{i}_j, \mathbf{x}_j) n^{-1/4} p(\varDelta_j \mathbf{i}, \varDelta_j \mathbf{x}) \to 0$$

in probability as $K \to \infty$, (8.30) is true for $K = \infty$. This yields (8.20). Then, the proof of (8.22) is similar. □

Remark 8.2

1. The authors in [56] give an extension with only six moments for the environment, and also to walks with long jumps. They place the result in the framework of Harris criterium for disorder relevance, and in [57] prove, with an elegant approach based on Fourth Moment Theorem, a log-normal limit for short range jumps in space dimension $d = 2$ with $\beta_n = \mathcal{O}(\ln^{1/2} n)$.
2. Heavy tails bring us back to the question discussed in Sect. 6.4. Recently [98] gave a glimpse of a bigger picture. Assuming $\mathbb{P}(\omega(t, x) > r) \asymp r^{-\alpha}$ and taking $\beta_n = n^{-\gamma}$ (for positive α, γ), what are the values of the exponents $\chi^{\|}, \chi^{\perp}$? This is a more ambitious question bridging many different regimes. Section 6.4 above deals with the case $\chi^{\perp} = 1$, and the above paragraph with the case $\chi^{\perp} = 1/2$. Results and conjectures are summarized in Fig. 1 in [98]. □

For the polymer measure, [6] gives further results: for the transition probabilities of the point-to-line measure,

$$\frac{n^{1/2}}{2} P_n^{\beta_n, \omega}(S_{nt} = n^{1/2}y | S_{ns} = n^{1/2}x) \xrightarrow{\text{law}} \frac{\mathscr{Z}_\beta(s, x; t, y)\mathscr{Z}_\beta(t, y; 1, *)}{\mathscr{Z}_\beta(s, x; 1, *)}. \tag{8.31}$$

The above convergence, as well as in Theorem 8.1, holds at the process level: [6] complete the marginal distribution convergence we sketched above with tightness of the time-space process. Then, after diffusive scaling, the discrete polymer in the intermediate disorder regime $\beta = \mathcal{O}(n^{-1/4})$ converges to the CRP.

Remark 8.3 We can sketch an explanation for Proposition 8.1, which claims that double-sided Brownian motion is invariant for KPZ equation.

We follow the approach of Flores et al. [179]. Consider stationary Log-Gamma polymer with parameter μ (this is achieved with boundary conditions $\mu/2$ placed at time $-\infty$): Using the notations of the previous chapter, $i, j \in \mathbb{Z}$, $Y_{i,j}^{-1} \sim \Gamma(\mu)$, $U_{i,j} = Z_{i,j}/Z_{i,j-1}$ and $U_{i,j} = Z_{i,j}/Z_{i-1,j}$ are inverse $\Gamma(\mu/2)$ with independence along down-right paths. We rotate by 45°: we use the convention

$$n = i + j, \quad x = i - j.$$

Assume $n = 2m$ even. Then,

$$h(n, x) \stackrel{\text{def}}{=} \ln Z_{i,j}/Z_{m,m} = \mathfrak{S}_x^{(n)},$$

with $\mathfrak{S}^{(n)}$ a double-sided random walk with increment $X_k^n = \ln U_{k,n-k}/V_{k+1,n-k-1}$,

$$\mathfrak{S}_x^{(n)} = \begin{cases} X_1^n + \ldots + X_j^n & x \geq 1, \\ 0, & x = 0 \\ -X_{-1}^n - \ldots - X_{j+1}^n & x \leq -1. \end{cases}$$

With $P_{n,\mu}$ denoting the n-steps log-gamma polymer law, stationarity writes

$$\sum_x e^{\mathfrak{S}^{(0)}(x)} P_{n,\mu}(S_n = y | S_0 = x) = e^{\mathfrak{S}^{(n)}(y) + \text{Shift}}, \tag{8.32}$$

where $\text{Shift} = \ln \sum_x e^{\mathfrak{S}^{(0)}(x)} P_{n,\mu}(S_n = 0 | S_0 = x)$. Corresponding to $\beta_n \simeq n^{-1/4}$, take

$$\mu = 2\sqrt{n}/\beta$$

and take limit $n \to \infty$ in (8.32): The counterpart of (8.31) is that the log-gamma polymer transitions converge to the KPZ as a process in the space variable, so we expect to get in the limit to have the following equality in law

$$\int_{\mathbb{R}} e^{\sqrt{2}\beta B(x)} P_{KPZ,\beta}(S_t = y | S_0 = x) \stackrel{\text{law}}{=} e^{\sqrt{2}\beta B(y) + \text{Shift}},$$

Since white noise is invariant by $(t, y) \mapsto (-t, y)$, we get also reversibility. $\qquad\square$

8.5　Fluctuations and Universality

Exponents For the CRP, the longitudinal exponent is $\chi^{\|} = 0$ since the normalized partition function converges to a nonzero limit. Since the polymer converges with a diffusive scaling to a density, the transverse exponent is $\chi^{\perp} = 1/2$ and there is no localization. The disorder for $\beta = \mathcal{O}(n^{-1/4})$ is not strong enough to place the model really far from Edwards-Wilkinson universality class [105]. This may look disappointing at first sight.

　　Looking back at (8.22) from a different perspective, we see that the logarithm of the P2P partition function takes the form of a stationary process minus a parabola. Such structure was first encountered for the polynuclear growth (PNG) droplet [195]. There, the stationary process was Airy$_2$, which has Tracy-Widom $F_{\text{GUE}} \equiv F_2$

as one-dimensional marginal distribution. Here, things are not so simple. A remarkable breakthrough was done by Amir et al. in [13] by giving a complete analytical description of the one-point probability distribution. Independently, Sasamoto and Spohn [206, 207] obtain a formula for the one-point marginal of A_β, and how it scales in β.

Recall A_β from (8.16). By (8.9), $A_\beta(1,x)$ is equal to

$$\ln\left(\sum_{k=0}^{\infty}\beta^k\int_{\Delta_k}\int_{\mathbb{R}^k}\frac{g(1-t_k,x-x_k)}{g(1,x)}\prod_{j=1}^{k}g(t_j-t_{j-1},x_j-x_{j-1})\dot{\mathscr{W}}(t_j,x_j)dt_jdx_j\right)$$

It has the following asymptotics.

Theorem 8.2 ([13]) *For $\beta > 0$ and $x \in \mathbb{R}$,*

$$\mathbb{P}\big(A_\beta(1,x)+\beta^4/4! \le \beta^{4/3}s\big) \longrightarrow F_{\text{GUE}}\big(2^{1/3}s\big), \quad \text{as } \beta \to \infty,$$

and

$$\beta^{-1}A_\beta(1,x) \quad \text{converges in law to} \quad \mathscr{N}\big(0,\pi^{1/2}/2\big), \quad \text{as } \beta \to 0^+.$$

The marginal distribution of $A_\beta(1,x)$ is called the crossover distribution. The name is easily understood, as the bridge from the weak disorder regime of small β with Gaussian fluctuations to the strong disorder regime of large β with Tracy-Widom universal fluctuations.[5] The scaling $\beta_n = \mathcal{O}(n^{-1/4})$ allows to capture the onset of the KPZ regime. A wider range $\beta_n = \beta n^{-1/\alpha}, \alpha \in (0,1/4]$ for the scaling of the temperature is studied in [54] at a non-rigorous level. It was taken up and formulated as a conjecture in [4], and later proved in [178] for the related model of the semi-discrete polymer of O'Connell-Yor at equilibrium.

Let us give a consequence of the Theorem.

Corollary 8.1 (Weak Universality) *Recall β_n from (8.19). When $n \to \infty$ first and then $\beta \to \infty$,*

$$\frac{\ln Z_{n,0}(\omega,\beta_n)-a_n}{2^{1/3}\beta^{4/3}} \xrightarrow{\text{law}} F_{\text{GUE}},$$

with $a_n = n\lambda(\beta_n)-(1/2)\ln(\pi n/2)-2\beta^4/3$.

---

[5]It is still a conjecture [13] if the process $A_\beta(1,\cdot)$ with a suitable renormalization converges to the Airy$_2$ process. Picture: Airy$_2(x) - x^2$. In [85] it is shown that the maximum value of Airy$_2$-process minus a parabola is $F_{\text{GOE}} \equiv F_1$-distributed. The authors in [177] obtain a formula for the joint density of the value and the location of the maximum. In physics literature, there even exists a formula for the joint density for the (related) maximum of N non-intersecting Brownian bridges [208].

This shows that, for large β, both KPZ equation, CRP and polymer model with $\beta_n = c\beta n^{-1/4}$ are in the Kardar-Parisi-Zhang universality class [83].

In the Edwards-Wilkinson universality class, time-space-fluctuations scale like 4:2:1 with Gaussian limits, though in KPZ universality class, time-space-fluctuations scale like 3:2:1 with Tracy-Widom limit laws.

The above result is referred to as "weak" since we need to take a vanishing inverse temperature first, and then to compensate by letting $\beta \to \infty$. It should be compared with the conjecture (see [43] for the five moment assumption):

Conjecture 8.1 (Strong Universality) For the discrete random polymer with i.i.d. environment ω having five finite moments,

$$\frac{\ln Z_{n,0}(\omega, \beta) - nC(\beta)}{D(\beta)n^{1/3}} \xrightarrow{\text{law}} F_{\text{GUE}},$$

for some constants C, D.

Fluctuations statistics are affected by the initial data. There are six fundamental geometries for the initial profile, leading to different one-point and multi-point limit distributions, different Tracy-Widom laws and Airy processes (see e.g., [83], Sect. 1.2.5).

8.5.1 Additional Remarks

For more details on the relation of KPZ with Interacting Particle Systems, growth models, we refer to the nice review articles [108, 109, 146]. We simply mention a few points:

- For the simple exclusion process, the height function has a simple interpretation: it is proportional to the particle current, i.e., the algebraic number of particles which have crossed and edge
- KPZ universality class extends also to random walks in dynamic random environment [23, 24].
- Using that KPZ is the limit of simple exclusion process in a weakly asymmetric regime, the volume exponent for KPZ is known to be $\chi^{\|} = 1/3$ in the stationary case $z_0(x) = \exp B(x)$: Precisely, in [22] it is shown that, as $t \to \infty$,

$$\text{Var}(\mathcal{H}(t,0)) \asymp t^{2/3}.$$

  This also follows from the similar estimate in the log-gamma polymer [211] or the semi-discrete polymer [178].
- Occurrence of Tracy-Widom distributions as scaling limits for polymers was anticipated by similar limits for random matrices starting from [232] and for last passage percolation after the pioneering paper of Johansson [144].

Chapter 9
Variational Formulas

In this chapter we closely follow two recent papers [118] and [199] by Georgiou, Rassoul-Agha, Seppäläinen and Yilmaz. They introduce two types of variational formulas for the free energy:

- Cocycles are (random and inhomogeneous) additive functions on $\mathbb{N} \times \mathbb{Z}^d$. They enter variational formulas for the free energy such as (9.11) below. Such formulas originate in the PhD thesis of Rosenbluth [203] building on ideas of stochastic homogenization.
- Gibbs variational formulas of Sect. 9.6, with a balance between energy and entropy. This is a more familiar route in statistical mechanics, implemented for disordered systems with increasing complexity in many papers including [41, 67, 209].

A common feature besides the two types of quenched variational formulas is that the medium being frozen, we penalize environments which have unlikely space averages. The ergodic theorem for cocycles, Theorem 9.6 below, show that cocycles provide natural penalizations which can be used to obtain quenched estimates.

Notations:
In this chapter, we take $H_n(S) = \sum_{i=0}^{n-1} \omega(i, S_i), H_{m,n} = \sum_{i=m-1}^{n-1} \omega(i, S_i)$.
Time-space coordinates will be denoted by: $\mathfrak{x} = (t, x), \mathfrak{y} = (s, y), \mathfrak{z} \in \mathbb{Z}^{1+d}$, with coordinates $\mathfrak{x}_0 = t, \mathfrak{x}_i = x_i (i = 1, \ldots d)$. In these coordinates, the path is denoted by $\mathfrak{s}_n = (n, S_n)$, and the origin $\mathfrak{o} = (0, 0)$
$\mathcal{N} = \{\pm e_i; i \leq d\}$ are the possible transverse jumps under P.
The open ball in $|\cdot|_1$-norm is $\mathscr{B}_1 = \{x \in \mathbb{R}^d : |x|_1 < 1\}$, and the closed ball by $\bar{\mathscr{B}}_1$.

© Springer International Publishing AG 2017
F. Comets, *Directed Polymers in Random Environments*, Lecture Notes
in Mathematics 2175, DOI 10.1007/978-3-319-50487-2_9

9.1 Free Energy Revised: A More Detailed Account

We will deal with many partition functions. Besides the n-step quenched "point-to-level" (level means line or hyperplane) partition function

$$Z_n = Z_n(\omega; \beta) = P[e^{\beta H_n(S)}] ,$$

we consider the tilted version

$$Z_n(h) = Z_n(\omega; \beta, h) = P[e^{\beta H_n(S) + h \cdot S_n}], \tag{9.1}$$

with an external field $h \in \mathbb{R}^d$. The n-step quenched point-to-point partition function is for $x \in \mathbb{Z}^d$,

$$Z_{n,x} = Z_{n,x}(\omega; \beta) = P[e^{\beta H_n(S)}; S_n = x].$$

Let $L_n = \{x \in \mathbb{Z}^d : P(S_n = x) > 0\}$ the set of attainable sites in n steps. To take limits of point-to-point quantities we specify lattice points $\hat{x}_n(\xi)$ that approximate $n\xi$ for $\xi \in \bar{\mathcal{B}}_1$: Fix $\hat{x}_n(\xi) \in L_n$ with $\|\hat{x}_n(\xi) - n\xi\|_\infty \leq d$.

Theorem 9.1 *Assume* $\mathbb{P}[|\omega|^p] < \infty$ *for some* $p > d$.

(a) *The nonrandom limit*

$$p_{\mathrm{pl}}(\beta) = p(\beta) = \lim_{n \to \infty} n^{-1} \ln Z_n(\omega; \beta)$$

exists $\mathbb{P}$-*a.s. and is finite. The same statement holds for*

$$p_{\mathrm{pl}}(\beta, h) = \lim_{n \to \infty} n^{-1} \ln Z_n(\omega; \beta, h) . \tag{9.2}$$

(b) *There exists an event of full* $\mathbb{P}$-*measure on which, for all* $\xi \in \bar{\mathcal{B}}_1$ *and any choices made in the definition of* $\hat{x}_n(\xi)$, *the limit*

$$p_{\mathrm{pp}}(\beta, \xi) = \lim_{n \to \infty} n^{-1} \ln Z_{n,\hat{x}_n(\xi)}(\omega; \beta) \tag{9.3}$$

exists. It is a nonrandom, concave and continuous function on all of $\bar{\mathcal{B}}_1$. *(The limit is independent of the particular choice of the* $\hat{x}^{(n)}(\xi)$ *above.) We have the almost sure identity*

$$p_{\mathrm{pl}}(\beta) = \sup_{\xi \in \bar{\mathcal{B}}_1 \cap \mathbb{Q}^d} p_{\mathrm{pp}}(\beta, \xi) = \sup_{\xi \in \bar{\mathcal{B}}_1} p_{\mathrm{pp}}(\beta, \xi) \tag{9.4}$$

The limits p_{pl} [resp., p_{pp}] are called point-to-level [resp., directional] free energy.

A sequence $(y_n)_n$ is called an admissible path if each finite subsequence is the path of the simple random walk with positive probability, i.e., for all m, $P(S_i = y_i, i \leq m) > 0$.

Proof Statements in (a) are covered by previous arguments (except for less integrability assumptions on ω). We only prove (b):

Step 1: For $\mathbb{P}$-a.e. ω and simultaneously for all $\xi \in \bar{\mathcal{B}}_1 \cap \mathbb{Q}^d$, for all admissible path $\{y_n(\xi)\}_{n \geq 0}$ such that for some integer k, $y_{mk}(\xi) = mk\xi$ for all $m \geq 0$, the limit

$$p_{pp}(\beta, \xi) = \lim_{n \to \infty} n^{-1} \ln Z_{n, y_n(\xi)}(\omega; \beta) \tag{9.5}$$

exists, and does not depend on the path $\{y_n(\xi)\}$ subject to the above condition. Fix $\xi \in \bar{\mathcal{B}}_1 \cap \mathbb{Q}^d$ and k such that $k\xi$ has integer coordinates. Then, for $m, n \geq 0$,

$$
\begin{aligned}
Z_{(n+m)k, (n+m)k\xi}(\omega; \beta) &= P\left[e^{\beta H_{k(n+m)}(S)}; S_{(n+m)k} = (n+m)k\xi\right] \\
&\geq P\left[e^{\beta H_{k(n+m)}(S)}; S_{mk} = mk\xi, S_{(n+m)k} = (n+m)k\xi\right] \\
&\overset{\text{Markov}}{=} P\left[e^{\beta H_{km}(S)}; S_{mk} = mk\xi\right] \\
&\quad \times P\left[e^{\beta H_{km, k(n+m)}(S)}; S_{(n+m)k} = (n+m)k\xi \big| S_{mk} = mk\xi\right] \\
&= Z_{mk, mk\xi}(\omega; \beta) Z_{nk, nk\xi}(\theta_{mk, mk\xi}\omega; \beta)
\end{aligned}
$$

[Compare with (2.3) and (2.8)]. Hence the doubly-indexed random sequence

$$X_{m, m+n}(\omega) = \ln Z_{nk, nk\xi}(\theta_{mk, mk\xi}\omega; \beta)$$

is superadditive,

$$X_{0,m} + X_{m, m+n} \leq X_{0, m+n},$$

and integrable. By Kingman's subadditive ergodic theorem (see Theorem A.1 in Appendix), we have a.s. convergence

$$\lim_{n \to \infty} (nk)^{-1} \ln Z_{nk, nk\xi}(\omega; \beta) = \sup_n (nk)^{-1} \mathbb{P} \ln Z_{nk, nk\xi}(\omega; \beta) =: p_{pp}(\beta, \xi).$$

The limit does not depend on k, since for k_1, k_2 as above, we have convergence along multiples of the product $k_1 k_2$. We now extend the above limit from multiples of k to the full sequence. Let n, m with $mk \leq n < (m+1)k$. Since $y.(\xi)$ is an admissible path, $|y_n(\xi) - y_{mk}(\xi)|_1 \leq k$, and we introduce

$$A_k(\omega) = \beta \max\left\{|H_{i,j}(s)|; 0 \leq i \leq j \leq k, s = (s_i)_{i=0}^k \text{ admissible}\right\},$$

which is integrable since the maximum is over a finite set. We have, with n, m as above,

$$-\big(k\ln(2d) + A_k(\theta_{mk,mk\xi}\omega)\big) \le \ln Z_{n,y_n(\xi)}(\omega;\beta) - \ln Z_{mk,mk\xi}(\omega;\beta) ,$$

$$\ln Z_{n,y_n(\xi)}(\omega;\beta) - \ln Z_{(m+1)k,(m+1)k\xi}(\omega;\beta) \le k\ln(2d) + A_k(\theta_{mk,mk\xi}\omega) .$$

We now argue that $\lim_m A_k(\theta_{mk,mk\xi}\omega)/m = 0$ a.s., which shows convergence of the full sequence. Indeed, as $m \to \infty$,

$$\frac{A_k(\theta_{mk,mk\xi}\omega)}{m} = \frac{1}{m}A_k(\omega) + \frac{1}{m}\sum_{\ell=1}^{m-1}\big(A_k(\theta_{(m+1)k,(m+1)k\xi}\omega) - A_k(\theta_{mk,mk\xi}\omega)\big)$$

$$\longrightarrow 0 + \mathbb{P}[A_k(\theta_{k,k\xi}\omega) - A_k(\omega)] = 0,$$

by the pointwise ergodic theorem. This proves the a.s. convergence in (9.5) for a fixed ξ. The full statement with simultaneous convergence for all $\xi \in \bar{\mathscr{B}}_1 \bigcap \mathbb{Q}^d$ follows by taking a countable intersection.

Step 2: Deduce the existence of the limit p_{pp} in (9.5) for irrational velocities ξ, on the event of full measure (by step 1) where the limit (9.5) exists for all rational ξ. See [197, Theorem 2.2].

Step 3: The resulting function $p_{pp}(\beta, \cdot)$ is continuous on $\bar{\mathscr{B}}_1$. Hence, the second equality in (9.4) holds.

Step 4: Concavity on rationals: For $\xi, \xi' \in \bar{\mathscr{B}}_1 \bigcap \mathbb{Q}^d$ and $\lambda \in [0, 1]\bigcap \mathbb{Q}$,

$$p_{pp}(\beta, \lambda\xi + (1-\lambda)\xi') \ge \lambda p_{pp}(\beta, \xi) + (1-\lambda)p_{pp}(\beta, \xi')$$

Indeed, pick an integer k such that $\lambda k, \lambda k\xi, (1-\lambda)k\xi'$ have integer coordinates. Then, by Markov inequality,

$$Z_{mk,m(\lambda k\xi+(1-\lambda)k\xi')}(\omega, \beta) \ge Z_{m\lambda k, m\lambda k\xi}(\omega, \beta)Z_{m(1-\lambda)k,m(1-\lambda)k\xi'}$$
$$\times (\theta_{mk,m(\lambda k\xi+(1-\lambda)k\xi')}\omega, \beta).$$

Taking logarithm and dividing by mk, as $m \to \infty$ we have convergence (in probability for the right-hand side) to the desired inequality.

Final step: By continuity, $p_{pp}(\beta, \cdot)$ is concave on $\bar{\mathscr{B}}_1$. $\square$

Remark 9.1 Convergence in (9.3) also holds in L^1-norm. See Remark 5.3 in [118]. Or use concentration inequalities, see e.g. [197, Lemma 8.1].

9.2 Large Deviation Principle for the Polymer Endpoint

Following an approach of Varadhan [233] for random walks in random environment, Carmona and Hu have shown a quenched large deviation principle for the polymer endpoint [59].

Theorem 9.2 ([59]) *There exists a deterministic function $\mathscr{I}_\beta : \mathbb{R}^d \to [0, +\infty]$, such that, $\mathbb{P}$-a.s., the law of S_n/n under the polymer measure $P_n^{\beta,\omega}$ obeys a large deviation principle with good rate $\mathscr{I}_\beta$. This means that there exists a set of environment Ω_0 with full $\mathbb{P}$-probability such that for all $\omega \in \Omega_0$, we have the following two inequalities: for all closed sets F of $\mathbb{R}^d$,*

$$\limsup_{n \to \infty} n^{-1} \ln P_n^{\beta,\omega}(S_n/n \in F) \le -\inf\{\mathscr{I}_\beta(\xi); \xi \in F\},$$

and for all open sets G of $\mathbb{R}^d$,

$$\liminf_{n \to \infty} n^{-1} \ln P_n^{\beta,\omega}(S_n/n \in G) \ge -\inf\{\mathscr{I}_\beta(\xi); \xi \in G\}.$$

The function $\mathscr{I}_\beta$ is convex and finite on $\bar{\mathscr{B}}_1$, infinite outside this set. It is invariant under the isometries of the lattice—and then $\mathscr{I}_\beta(0) = 0$—, and is bounded by

$$0 \le \mathscr{I}_\beta(\xi) \le \ln(2d) + p(\beta) - \beta\lambda'(0).$$

If $\xi = \pm e_i$, $\{S_n = n\xi\}$ contains only one path, with all jumps equal to ξ. In that case, the law of large numbers along this path yields the explicit value

$$I_\beta(\pm e_i) = \ln(2d) + p(\beta) - \beta\lambda'(0)$$

In general, using the results of the previous section, we can identify the rate function,

$$\mathscr{I}_\beta(\xi) = p(\beta) - p_{pp}(\beta, \xi). \tag{9.6}$$

Exercise 9.1 The aim of this exercise is to prove that, in dimension larger or equal to 3, the rate function of the polymer is equal to the one of the simple random walk in a neighborhood of the origin, provided that β is small enough. This behavior is reminiscent of a similar one for random walks in random environments [242] or in random potential [245].

Let $\alpha = (\alpha_1, \ldots, \alpha_d) \in \mathbb{R}^d$, $\cosh u = (e^u + e^{-u})/2$, and

$$\phi(\alpha) = \ln \left(\frac{1}{d} \sum_{i=1}^{d} \cosh \alpha_i \right) = \ln P[e^{\alpha \cdot S_1}].$$

Denote by $P^{(\alpha)}$ the law of the (asymmetric) random walk with i.i.d. jumps with law

$$P^\alpha (S_n - S_{n-1} = e) = \frac{\exp\{\alpha \cdot e\}}{2 \sum_{i=1}^{d} \cosh \alpha_i}$$

for all $e \in \mathcal{N}$. Define

$$W_n^{(\alpha)} = P \left[\exp \left(\beta H_n(S) - n\lambda(\beta) + \alpha \cdot S_n - n\phi(\alpha) \right) \right]$$

1. Check that $W_n^{(\alpha)}$ is a positive martingale, and that

$$W_n^{(\alpha)} = P^{(\alpha)} \left[\exp \left(\beta H_n(S) - n\lambda(\beta) \right) \right]$$

2. Find the constant $C = C_\beta$ such that

$$\mathbb{P}[(W_n^{(\alpha)})^2] = (P^{(\alpha)})^{\otimes 2} \left[e^{CN_n} \right].$$

3. Deduce that for $d \geq 3$, there exists $\beta_0(\alpha) > 0$ such that for $|\beta| < \beta_0(\alpha)$, $W_n^{(\alpha)}$ is bounded in L^2, and that, for $|\beta| < \beta_0(\alpha) \wedge \beta_0(0)$, we have

$$\lim_{n \to \infty} \frac{1}{n} \ln P_n^{\beta, \omega}[\exp(\alpha \cdot S_n)] = \phi(\alpha) \qquad \text{a.s.}$$

4. Conclude that, for $d \geq 3$ and $\epsilon \in (0, 1)$, we can take $\beta > 0$ small enough such that the rate function $\mathcal{I}_\beta$ coincide with that of the non-random model in the ϵ-neighborhood of 0,

$$\mathcal{I}_\beta(\xi) = \sup\{\alpha \cdot \xi - \phi(\alpha)\}, \qquad \|\xi\|_1 < \epsilon.$$

Open Problem 9.3 *It is expected that, for all values of β, the rate function $\mathcal{I}_\beta$ is strictly convex on its domain. Equivalently, the point-to-point free energy is expected to be strictly concave. This property is the curvature assumption which is assumed in many works, e.g. [16].*

9.3 Cocycle Variational Formula for the Point-to-Level Case

Since the ω's are i.i.d. under $\mathbb{P}$, the shift $\theta_{k,y}$ is measure-preserving bijection on $(\Omega, \mathscr{A}, \mathbb{P})$.

Definition 9.1 (Stationary Cocycles)

(i) A measurable function $f : \Omega \times \mathscr{N} \to \mathbb{R}$ is a stationary cocycle if

$$\sum_{i=0}^{n-1} f(\theta_{i,x_i}\omega, \Delta_i x) = \sum_{i=0}^{n-1} f(\theta_{i,y_i}\omega, \Delta_i y) \qquad \text{with } \Delta_i x = x_{i+1} - x_i,$$

for all admissible paths x, y from $\mathfrak{o} = (0,0)$ to $\mathfrak{z} = (n,z)$. (Note that $\Delta_i x = \Delta x_{i+1}$ defined in Sect. 5.1.)

(ii) Then the common value is denoted by $F(\omega, \mathfrak{o}, \mathfrak{z})$, and F can be extended to a function $F : \Omega \times \mathscr{G}^{2+} \to \mathbb{R}$ by

$$F(\omega, \mathfrak{y}, \mathfrak{z}) = \sum_{i=m}^{n-1} f(\theta_{i,x_i}\omega, \Delta_i x), \qquad x \text{ admissible path from } \mathfrak{y} \text{ to } \mathfrak{z},$$

with $\mathfrak{y}_0 = m, \mathfrak{z}_0 = n$. The set $\mathscr{G}$ is the additive group generated by the time-space steps $(1, z), z \in \mathscr{N}$, i.e., the set of $(t, x) \in \mathbb{Z}^2$ with $t + x = 0$ modulo 2; the set $\mathscr{G}^{2+} = \{(\mathfrak{y}, \mathfrak{z}) : \mathfrak{y} \in \mathscr{G}, \mathfrak{z} - \mathfrak{y} \in \bigcup_{n \geq 0}\{(n, x); x \in L_n\}\}$. This extension is stationary, i.e.,

$$F(\omega, \mathfrak{z} + \mathfrak{x}, \mathfrak{z} + \mathfrak{y}) = F(\theta_{\mathfrak{z}}\omega, \mathfrak{x}, \mathfrak{y}), \tag{9.7}$$

and satisfies the usual definition of a cocycle as an additive function,

$$F(\omega, \mathfrak{x}, \mathfrak{y}) + F(\omega, \mathfrak{y}, \mathfrak{z}) = F(\omega, \mathfrak{x}, \mathfrak{z}). \tag{9.8}$$

The correspondence $f \mapsto F$ being one-to-one, with reciprocal $f(\omega, z) = F(\omega, \mathfrak{o}, (1, z))$, so we will use the name cocycle for both f defined by (i) and F by (ii)—i.e., satisfying (9.7) and (9.8).

If $\mathbb{P}|f(\omega, z)| < \infty$ we say the cocycle is integrable, and centered if moreover $\mathbb{P}f(\omega, z) = 0$ for all $z \in \mathscr{N}$.

Let $\mathscr{K}$ denote the space of stationary $L^1(\mathbb{P})$ cocycles, and $\mathscr{K}_0$ denote the subspace of centered stationary cocycles. We write $f \in \mathscr{K}$ or $F \in \mathscr{K}$ indifferently. The term cocycle is borrowed from differential forms terminology. One could also use the term conservative flow following vector fields terminology. The space $\mathscr{K}_0$ is the $L_1(\mathbb{P})$-closure of gradients $F(\omega, \mathfrak{x}, \mathfrak{y}) = \phi(\theta_{\mathfrak{y}}\omega) - \phi(\theta_{\mathfrak{x}}\omega)$, see [198, Lemma C.3].

Centering For $B \in \mathscr{K}$ there exists a vector $\mathfrak{h}(B) = (h_0(B), h(B)) \in \mathbb{R} \times \mathbb{R}^d$ such that

$$\mathbb{P}\big[B(\mathfrak{o}, \mathfrak{z})\big] = -\mathfrak{h}(B) \cdot \mathfrak{z}, \qquad \forall \mathfrak{z} \in \mathscr{G}. \tag{9.9}$$

(For brevity, we omit ω in the notation of $B(\omega, \mathfrak{x}, \mathfrak{y})$.) Existence of $\mathfrak{h}(B)$ follows because $c(\mathfrak{x}) = \mathbb{P}[B(\mathfrak{o}, \mathfrak{x})]$ is an additive function on the group $\mathscr{G}$ by (9.7) and (9.8). The linear system (9.9) looks overdetermined, with $2d$ equations for $d+1$ unknown. In fact, it has a unique solution

$$\begin{cases} h_i(B) = -\frac{1}{2}\big[\mathbb{P}B(\mathfrak{o}, (1, e_i)) - \mathbb{P}B(\mathfrak{o}, (1, -e_i))\big], \\ h_0(B) = -\frac{1}{2}\big[\mathbb{P}B(\mathfrak{o}, (1, e_i)) + \mathbb{P}B(\mathfrak{o}, (1, -e_i))\big], \end{cases} \quad i = 1, \ldots d.$$

Indeed, in that case, we have

$$\mathbb{P}B(\mathfrak{o}, (1, \pm e_i)) = -\big[h_0(B) \pm h_i(B)\big],$$

which is (9.9). Then

$$F_B(\mathfrak{x}, \mathfrak{y}) = -B(\mathfrak{x}, \mathfrak{y}) - \mathfrak{h}(B) \cdot (\mathfrak{y} - \mathfrak{x}) \tag{9.10}$$

is a centered stationary $L^1(\mathbb{P})$ cocycle.

Theorem 9.4 *The limit in* (9.2) *has the variational representation:*

$$p_{\mathrm{pl}}(\beta, h) = \inf_{f \in \mathscr{K}_0} \mathbb{P}-\mathrm{ess} \sup_{\omega} \ln P\left[e^{\beta\omega(0,0)+h\cdot S_1+f(\omega,S_1)}\right]. \tag{9.11}$$

A minimizing $f \in \mathscr{K}_0$ exists for each β and each $h \in \mathbb{R}^d$.

With $\mathscr{G}_{0,\infty} = \sigma(\omega(t, x); t \geq 0, x \in \mathbb{Z}^d\}$, let

$\quad L^+$ be the set of $\mathscr{G}_{0,\infty}$-measurable functions on Ω which are $\mathbb{P}$-a.s. positive,

and

$\quad L^{++}$ be those which are $\mathbb{P}$-a.s. bounded away from 0.

Special cocycles are *logarithmic gradients*, i.e., cocycles of the form

$$F(\omega, \mathfrak{x}, \mathfrak{y}) = \ln\big[g(\theta_{\mathfrak{y}}\omega)/g(\theta_{\mathfrak{x}}\omega)\big]$$

with $g \in L^+$ or $g \in L^{++}$ in (9.11).

Theorem 9.5 ([199, Theorem 2.1]) *The limit in (9.2) has the variational representation:*

$$p_{pl}(\beta, h) = \inf_{g \in L^+} \mathbb{P}-\text{ess} \sup_{\omega} \ln P \left[e^{\beta \omega(0,0) + h \cdot S_1} \times \frac{g(\theta_{1, S_1} \omega)}{g(\omega)} \right] \tag{9.12}$$

$$= \inf_{g \in L^{++}} \mathbb{P}-\text{ess} \sup_{\omega} \ln P \left[e^{\beta \omega(0,0) + h \cdot S_1} \times \frac{g(\theta_{1, S_1} \omega)}{g(\omega)} \right]. \tag{9.13}$$

In this form, the above variational formula looks like as an infinite-dimensional version of the min-max variational formula for the Perron-Frobenius eigenvalue of a nonnegative matrix. See Appendix A.6, formula (A.19) and the conclusion of that section.

Proof We prove both theorems at the same time, following the lines of proof of Theorem 2.1 in [199] and [198, Theorem 2.3]. Introduce the notation $\mathfrak{s}_n = (n, S_n)$ for $n \geq 1$.

For $F \in \mathcal{K}_0$, denote

$$K(F) = \mathbb{P}-\text{ess} \sup_{\omega} \ln P \left[e^{\beta \omega(\mathfrak{o}) + h \cdot S_1 + F(\omega, \mathfrak{o}, \mathfrak{s}_1)} \right].$$

Step 1: Lower bound. Recall that $p_{pl}(\beta, h) < \infty$. For all $\gamma > p_{pl}(\beta, h)$, define for $n \geq 1$

$$h_n^\gamma = P \left[e^{\beta H_n + h \cdot S_n - n\gamma} \right] = Z_n(\beta, h) e^{-n\gamma}, \qquad h_0^\gamma = 1, \tag{9.14}$$

$$\text{and} \qquad g^\gamma = \sum_{n \geq 0} h_n^\gamma \geq 1, \tag{9.15}$$

which is a.s. convergent since

$$\lim_n n^{-1} \ln h_n^\gamma = p_{pl}(\beta, h) - \gamma < 0, \qquad \mathbb{P} - \text{a.s.}$$

We can write

$$g^\gamma(\omega) = 1 + \sum_{n \geq 1} h_n^\gamma$$

$$= 1 + \sum_{n \geq 1} P \left[e^{\beta \omega(0,0) + h \cdot S_1 - \gamma} h_{n-1}^\gamma(\theta_{1, S_1} \omega) \right]$$

$$= 1 + P \left[e^{\beta \omega(0,0) + h \cdot S_1 - \gamma} g^\gamma(\theta_{1, S_1} \omega) \right]$$

bringing the sum inside the expectation. Reorganizing the terms, we get

$$\gamma = \ln \left(\frac{e^\gamma}{g^\gamma(\omega)} + P\left[e^{\beta\omega(0,0)+h\cdot S_1} \frac{g^\gamma(\theta_{1,S_1}\omega)}{g^\gamma(\omega)} \right] \right)$$
$$> \ln \left(P\left[e^{\beta\omega(0,0)+h\cdot S_1} \frac{g^\gamma(\theta_{1,S_1}\omega)}{g^\gamma(\omega)} \right] \right).$$

This means that for the cocycle f^γ given as the logarithmic gradient $f^\gamma(\omega, z) = \ln g^\gamma(\theta_{1,z})/g^\gamma(\omega)$ with $g^\gamma \geq 1$, we have $\gamma \geq K(f^\gamma)$. Therefore,

$$p_{\mathrm{pl}}(\beta, h) = \inf_{\gamma > p_{\mathrm{pl}}(\beta,h)} \gamma$$
$$\geq \inf_{\gamma > p_{\mathrm{pl}}(\beta,h)} K(f^\gamma)$$
$$\geq \inf_{f \text{ log-gradient of } g \in L^{++}} K(f)$$
$$\geq \inf_{f \in \mathcal{K}_0} K(f), \tag{9.16}$$

yielding the lower bound.

Step 2: Upper bound. $p_{\mathrm{pl}}(\beta, h) \leq K(f)$ for $f \in \mathcal{K}_0$. It suffices to consider the case $K(f) < \infty$, otherwise there is nothing to prove. Recalling the set L_n of attainable sites in n steps, we have by Markov property

$$P\left[e^{\beta H_n + h\cdot S_n + \sum_{i=0}^{n-1} f(\theta_{i,S_i}\omega, \Delta_i S)} \right]$$
$$= \sum_{x \in L_{n-1}} P\left[e^{\beta H_{n-1} + h\cdot S_{n-1} + \sum_{i=0}^{n-2} f(\theta_{i,S_i}\omega, \Delta_i S)}; S_{n-1} = x \right]$$
$$\times P\left[e^{\beta\omega(n-1,x)+h\cdot\Delta_{n-1}S+f(\theta_{n-1,x}\omega, \Delta_{n-1}S)} \big| S_{n-1} = x \right]$$
$$\leq P\left[e^{\beta H_{n-1}+h\cdot S_{n-1}+\sum_{i=0}^{n-2} f(\theta_{i,S_i}\omega, \Delta_i S)} \right] \times e^{K(f)}$$
$$\overset{\text{recursion}}{\leq} e^{nK(f)}.$$

Recall that $\sum_{i=0}^{n-1} f(\theta_{i,S_i}\omega, \Delta_i S) = F(\omega, \mathfrak{o}, \mathfrak{s}_n)$, $(\mathfrak{s}_n = (n, S_n))$, and note that the sum depends only on the ending point and not on the full path. We have

$$p_{\mathrm{pl}}(\beta, h) = \lim_{n\to\infty} n^{-1} \ln Z_n(\omega; \beta, h)$$
$$\leq \limsup_{n\to\infty} n^{-1} \ln P\left[e^{\beta H_n + h\cdot S_n + F(\omega, \mathfrak{o}, \mathfrak{s}_n)} \right]$$
$$+ \limsup_{n\to\infty} \max_{\mathfrak{s}_n : \mathfrak{s}_n \in L_n} \{ n^{-1} |F(\omega, \mathfrak{o}, \mathfrak{s}_n)| \} \tag{9.17}$$
$$\leq K(f) + \limsup_{n\to\infty} \max_{\mathfrak{s}_n : \mathfrak{s}_n \in L_n} \{ n^{-1} |F(\omega, \mathfrak{o}, \mathfrak{s}_n)| \},$$

using the previous bound. We now end the proof of the upper bound by arguing that the $\limsup_{n\to\infty}$ is zero. We check the following assumption:

$$\exists p > d, \exists F(\cdot, z) \in L^p(\Omega)(\forall z \in \mathcal{N}) : f(\omega, z) \leq \bar{F}(\omega, z). \tag{9.18}$$

Indeed, by definition of $K(f)$, we can take $\bar{F} = \beta|\omega(0,0)| + \ln(2d) + |h| + K(f)$ and any p. Under this condition the following ergodic theorem for cocycles has been recently proved.

Theorem 9.6 (Ergodic Theorem for Cocycles [117]) *Let F be a centered stationary cocycle with Property* (9.18) *for some $p > d$. Then,*

$$\lim_{n\to\infty} \max_{\mathfrak{s}_n : \mathfrak{s}_n \in L_n} \{n^{-1}|F(\omega, \mathfrak{o}, \mathfrak{s}_n)|\} = 0$$

The proof is given in [117], Theorem A.3 in the appendix, it will be omitted in these notes. Theorem 9.6 is needed in order to show (9.11). But, to complete the proof of (9.12), one can take the following subtle shortcut.

Remark 9.2 In the particular case where the centered cocycle f is the logarithmic gradient of a function bounded away from 0,

$$f(\omega, z) = \ln[g(\theta_{1,z}\omega)/g(\omega)], \qquad g : \Omega \to [a, +\infty) \text{ measurable,}$$

with $a > 0$, we have

$$F(\omega, \mathfrak{o}, \mathfrak{s}_n) = \ln g(\theta_{\mathfrak{s}_n}\omega) - \ln g(\omega) \geq \ln a - \ln g(\omega),$$

and then

$$\liminf_{n\to\infty} \min_{\mathfrak{s}_n : \mathfrak{s}_n \in L_n} \{n^{-1}F(\omega, \mathfrak{o}, \mathfrak{s}_n)\} \geq 0, \qquad \mathbb{P} - \text{a.s.}$$

Now we can improve the bound (9.17):

$$
\begin{aligned}
p_{\mathrm{pl}}(\beta, h) &\leq \limsup_{n\to\infty} n^{-1} \ln P\left[e^{\beta H_n + h \cdot S_n + F(\omega, \mathfrak{o}, \mathfrak{s}_n)}\right] \\
&\quad - \liminf_{n\to\infty} \min_{\mathfrak{s}_n : \mathfrak{s}_n \in L_n} \{n^{-1}F(\omega, \mathfrak{o}, \mathfrak{s}_n)\} \\
&\leq K(f),
\end{aligned}
\tag{9.19}
$$

completing the proof of (9.12).

Step 3: Existence of a minimizer. We now end by proving the existence of a minimizer in the variational formula (9.11). Let $f_i \in \mathcal{K}_0$ be a minimizing sequence, such that

$$P\left[e^{\beta\omega(0,0) + h \cdot S_1 + f_n(\omega, S_1)}\right] \leq \exp\{p_{\mathrm{pl}}(\beta, h) + 1/n\} \qquad \text{a.s.}$$

Hence, for all $z \in \mathcal{N}$,

$$f_n(\cdot, z) \leq |\beta\omega(0,0)| + \ln(2d) + p_{\mathrm{pl}}(\beta, h) + 1/n,$$

showing that $f_n^+(\cdot, z)$ is uniformly integrable. The f_n are centered by definition, so $\mathbb{P}f_n^-(\cdot, z) = \mathbb{P}f_n^+(\cdot, z)$ and $\mathbb{P}f_n^-(\cdot, z)$ is uniformly bounded. We use a (subtle) Lemma [154, Lemma 4.3], we do not reproduce the proof here, but the reader will find it there.

Lemma 9.1 *From a sequence $h_n \geq 0$ with*

$$\sup_n \mathbb{P}h_n \leq C,$$

we can extract a subsequence n_i and find a decomposition $h_{n_i} = \hat{h}_{n_i} + r_{n_i}$, with $\hat{h}_{n_i} \geq 0, r_{n_i} \geq 0$ and

- *$(\hat{h}_{n_i})_{i \geq 1}$ uniformly integrable,*
- *$r_{n_i} \to 0$ in probability as $i \to \infty$.*

Extract a common subsequence for all z, we obtain from the Lemma a decomposition

$$f_{n_i}^-(\cdot, z) = \hat{f}_{n_i}^-(\cdot, z) + r_{n_i}(\cdot, z)$$

with $\hat{f}_{n_i}^-(\cdot, z)$ uniformly integrable and $r_{n_i}(\cdot, z) \geq 0$ vanishing in probability. Extracting a further subsequence, $\tilde{f}_{n_i}(\cdot, z) = f_{n_i}^+(\cdot, z) - \hat{f}_{n_i}^-(\cdot, z)$ is weakly convergent in L^1 to some $\tilde{f}(\cdot, z)$. By Rudin [205, Theorem 3.12] $\tilde{f}(\cdot, z)$ is in the strong L^1-closure of the convex hull of $(\tilde{f}_{n_i}(\cdot, z); i \geq 1)$, i.e., it is the L^1-limit as $i \to \infty$ of some sequence of finite convex combinations $\tilde{g}_i(\cdot, z) = \sum_{k \geq i} \alpha_i(k) \tilde{f}_{n_k}(\cdot, z)$. Taking a subsequence we can assume the $\tilde{g}_i(\cdot, z) \to \tilde{f}(\cdot, z)$ $\mathbb{P}$-a.s. Since the f_n's are cocycles, this implies that $\tilde{f}(\cdot, z)$ is a cocycle. Also, from $r_{n_i}(\cdot, z) \geq 0$ we get $c(z) := \mathbb{P}\tilde{f}(\cdot, z) \geq 0$. Now, set $f(\cdot, z) = \tilde{f}(\cdot, z) - c(z)$ for all z. Then, $f \in \mathcal{K}_0$, and combining Jensen's inequality for the convex combination together with the minimizing sequence property, we get

$$P\left[e^{\beta\omega(0,0)+h\cdot S_1+\tilde{g}_i(\omega,S_1)-\sum_{k \geq i}\alpha_i(k)r_{n_k}(\cdot,S_1)}\right] \leq \exp\{p_{\mathrm{pl}}(\beta, h) + 1/n_i\} \qquad \text{a.s.}$$

Sending $i \to \infty$,

$$P\left[e^{\beta\omega(0,0)+h\cdot S_1+f(\omega,S_1)+c(S_1)}\right] \leq \exp\{p_{\mathrm{pl}}(\beta, h)\} \qquad \text{a.s.}$$

Hence we find at the same time that (i) $c(z) = 0$ for all z (otherwise, f would beat the bound (9.11), which is a contradiction), (ii) $f(\omega, z)$ is a minimizer of (9.11). □

The next definition and theorem offer a way to identify a minimizing F. Later we explain how Busemann functions provide minimizers that match this recipe.

Definition 9.2 A stationary L^1 cocycle B recovers the potential $V_0 : \Omega \to \mathbb{R}$ at temperature inverse β if

$$P\left[e^{\beta V_0(\omega) - B(\omega, o, s_1)}\right] = 1 \qquad \text{for } \mathbb{P} - \text{a.e. } \omega. \tag{9.20}$$

In our case we will consider only potential $V_0(\omega) = \omega(0,0)$, the environment at the origin.

Sometimes, the potential will depend in addition on the first step: $V_h(\omega, z) = \omega(0,0) + \beta^{-1}h \cdot z$. However, if B solves (9.20), then the formula $\bar{B}(\omega, o, \mathfrak{z}) = B(\omega, o, \mathfrak{z}) + h \cdot z, \mathfrak{z} = (1, \zeta)$, defines a new stationary cocycle which recovers the potential V_h, in the sense that

$$P\left[e^{\beta V_h(\omega, S_1) - \bar{B}(\omega, o, s_1)}\right] = 1 \qquad \text{for } \mathbb{P} - \text{a.e. } \omega.$$

Note also that, for B as in (9.20) and $\mathfrak{h} = (h_0, h) = \mathfrak{h}(B)$ from (9.9), the centered cocycle F_B,

$$F_B(\mathfrak{x}, \mathfrak{y}) = -B(\mathfrak{x}, \mathfrak{y}) - \mathfrak{h} \cdot (\mathfrak{y} - \mathfrak{x}) \quad \text{solves} \quad P\left[e^{\beta \omega(o) + h \cdot S_1 + F_B(\omega, o, s_1)}\right] = e^{-h_0} \; \mathbb{P} - \text{a.s.} \tag{9.21}$$

Theorem 9.7 *Suppose we have a stationary L^1 cocycle B that recovers the potential $\omega(0,0)$. Define $\mathfrak{h}(B) = (h_0(B), h(B))$ and $F = F_B$ as in (9.10)–(9.9). Then*

(i) $p_{pl}(\beta, h(B)) = -h_0(B)$.

(ii) F is a minimizer the variational formula in (9.11) for $p_{pl}(\beta, h(B))$. The essential supremum in (9.11) disappears and we have, for $\mathbb{P}$-a.e. ω,

$$p_{pl}(\beta, h(B)) = \ln P\left[e^{\beta V_0(\omega) + h(B) \cdot S_1 + F(\omega, (0,0), (1, S_1))}\right] = -h_0(B). \tag{9.22}$$

Correctors A cocycle that solves (9.11) without the essential supremum—that is, satisfies the first equality of (9.22)—is called a corrector. This term is used in an analogous manner in the homogenization literature [14, p. 468].

Proof of Theorem 9.7 From (9.21), with $F = F_B$ as in (9.10), we have

$$\ln P\left[e^{\beta \omega(o) + h(B) \cdot S_1 + F(\omega, o, s_1)}\right] = -h_0(B) \tag{9.23}$$

for $\mathbb{P}$-a.e. ω. By the Markov and cocycle properties, it implies that

$$
\begin{aligned}
\ln P\left[e^{\beta H_n + h(B)\cdot S_n + F(\omega,\mathfrak{o},\mathfrak{s}_n)}\right] &= \ln P\left[e^{\beta H_{n-1} + h(B)\cdot S_{n-1} + F(\omega,\mathfrak{o},\mathfrak{s}_{n-1})}\right.\\
&\qquad \times \left. P\left[e^{\beta\omega(\mathfrak{s}_{n-1}) + h(B)\cdot\Delta_{n-1}S + F(\omega,\mathfrak{s}_{n-1},\mathfrak{s}_n)}\middle| S_i, i\le n-1\right]\right]\\
&\overset{(9.23)}{=} \ln P\left[e^{\beta H_{n-1} + h(B)\cdot S_{n-1} + F(\omega,\mathfrak{o},\mathfrak{s}_{n-1})}\right] - h_0(B)\\
&= -n h_0(B),
\end{aligned}
$$

by iterating. The recovery assumption yields a bound of the form (9.18) with $\bar{F}(\omega, z) = -\omega(0,0) + C$, so the ergodic theorem for cocycles (Theorem 9.6) applies, and gives $F(\omega, \mathfrak{o}, \mathfrak{s}_n) = o(n)$ uniformly in $\mathfrak{s}_n$ with $s_n \in L_n$. Then, the last equality implies, in the limit $n \to \infty$,

$$p_{\mathrm{pl}}(\beta, h(B)) = -h_0(B).$$

This is (i), but (ii) then follows using (9.23). See [118, Theorem 3.4] for more details. $\qquad\qquad\square$

Example 9.1 (Directed Polymer in Weak Disorder) Fix $\beta, h \in \mathbb{R}^d$, and assume that the martingale $W_n = Z_n(\omega; \beta, h)e^{-n(\lambda(\beta) + \kappa(h))}$ (with $\kappa(h) = \ln P[e^{h\cdot S_1}]$) is uniformly integrable.

$\triangleright$ In that case, the limit W_∞ is a.s. positive. This gives us the limiting point-to-level free energy:

$$
\begin{aligned}
p_{\mathrm{pl}}(\beta, h) &= \lim_{n\to\infty} \frac{1}{n} \ln P\left[e^{\beta H_n + h\cdot S_n}\right]\\
&= \lim_{n\to\infty} \frac{1}{n} \ln W_n + \left(\lambda(\beta) + \kappa(h)\right)\\
&= \lambda(\beta) + \kappa(h).
\end{aligned}
\tag{9.24}
$$

Decomposition according to the first step (Markov property) and a passage to the limit gives

$$W_\infty(\omega) = e^{-(\lambda(\beta) + \kappa(h))} P\left[e^{\beta\omega(0,0) + h\cdot S_1} W_\infty(\theta_{1,S_1}\omega)\right].$$

Together with (9.24), this gives

$$p_{\mathrm{pl}}(\beta, h) = \ln P\left[e^{\beta\omega(0,0) + h\cdot S_1 + \beta f(\omega, S_1)}\right], \tag{9.25}$$

with f associated to the gradient

$$F(\omega, \mathfrak{x}, \mathfrak{y}) = \ln W_\infty(\theta_{\mathfrak{y}}) - \ln W_\infty(\theta_{\mathfrak{x}}). \tag{9.26}$$

It remains to verify that $f(\omega, z)$ is integrable for all $z \in \mathcal{N}$, since this implies that it is mean-zero. Equation (9.25) gives an upper bound that shows

$$\mathbb{P}\big[f(\omega, z)^{+}\big] < \infty, \qquad z \in \mathcal{N}. \tag{9.27}$$

We argue indirectly that also $\mathbb{P}\big[f(\omega, z)^{-}\big] < \infty$. By stationarity of the medium, the first limit in probability below holds in probability:

$$0 = \lim_{n \to \infty} \big((n\beta)^{-1} \ln W_{\infty}(\theta_{n,nz}\omega) - (n\beta)^{-1} \ln W_{\infty}(\omega)\big)$$

$$= \lim_{n \to \infty} n^{-1} \sum_{k=0}^{n} f(\theta_{k,kz}\omega, z).$$

Since $\mathbb{P}[f(\omega, z)^{+}] < \infty$ and by the ergodic theorem, assuming $\mathbb{P}[f(\omega, z)^{-}] = \infty$ would force the limit above to be $-\infty$, a contradiction. So $f(\omega, z) \in L^{1}$.

To summarize, (9.25) shows that the centered cocycle F satisfies (9.22) for $V(\omega, z) = \omega_0 + h \cdot z$ for this particular value (β, h). F is the corrector given in (9.10), from the cocycle B given by

$$B\big(\omega, (m, x), (n, y)\big) = p_{\mathrm{pl}}(\beta, h)(n - m) - h \cdot (x - y) - F\big(\omega, (m, x), (n, y)\big).$$

Cocycle B has $\mathfrak{h}(B) = (-p_{\mathrm{pl}}(\beta, h), h) \in \mathbb{R}^{1+d}$ and recovers the potential $V(\omega, z) + h_0$.

$\triangleleft$

Remark 9.3 (Existence of Minimizers) At weak disorder, the minimizer in (9.12) exists and is given by $g = W_{\infty}$.

Now, it is shown [199, Sect. 2.3] that (9.12) does not always have minimizers at strong disorder.

Remark 9.4 (Annealed Variational Formula) It is shown [199, Theorem 2.4, for a short proof] that the annealed free energy—i.e., the left-hand side of (9.28) below— is given by the variational formula,

$$\lambda(\beta) + \kappa(h) = \inf_{g \in L^1 \cap L^+} \mathbb{P}-\operatorname{ess\,sup}_{\omega} \ln E\left[e^{\beta\omega(0,0)+h \cdot S_1} \frac{g(\theta_{1,S_1}\omega)}{g(\omega)}\right] \tag{9.28}$$

$$= \inf_{g \in L^1 \cap L^{++}} \mathbb{P}-\operatorname{ess\,sup}_{\omega} \ln E\left[e^{\beta\omega(0,0)+h \cdot S_1} \frac{g(\theta_{1,S_1}\omega)}{g(\omega)}\right] \tag{9.29}$$

to be compared with (9.11), (9.12).

At weak disorder, $g = W_{\infty}$ is minimizing. In the directed case, we can compute the value of $n^{-1} \ln \mathbb{P}Z_n(\beta, h) = \lambda(\beta) + \kappa(h)$, and then the value of the annealed free energy. But the point is that, even for non-directed models, the RHS of (9.28) is equal to the annealed free energy $\lim_{n \to \infty} n^{-1} \ln \mathbb{P}Z_n(\beta, h)$, which is then not explicit anymore. (See [199, Sect. 2.3].)

Proof Proof of upper bounds in (9.28): denoting by $k(g)$ the essential supremum, we have for all such g,

$$g(\omega) \geq E\left[e^{\beta\omega(0,0)+h\cdot S_1 - k(g)} \times g(\theta_{1,S_1}\omega)\right].$$

Since g is $\mathscr{G}_{0,\infty}$-measurable, we can compute by independence

$$\mathbb{P}g(\omega) \geq E\left[\mathbb{P}(e^{\beta\omega(0,0)+h\cdot S_1 - k(g)}) \times \mathbb{P}(g(\theta_{1,S_1}\omega))\right]$$
$$= e^{\lambda(\beta)+\kappa(h)-k(g)}\mathbb{P}g \qquad (9.30)$$

by stationarity, so $\lambda(\beta) + \kappa(h) \leq k(g)$ for all $g \in L^1 \cap L^+$.

Proof of lower bounds in (9.28): Let $\gamma > \lambda(\beta) + \kappa(h)$ and g^γ from (9.15). Then, $g^\gamma \in L^1 \cap L^{++}$, and

$$\mathbb{P}g^\gamma = \left(1 - \exp\{\lambda(\beta) + \kappa(h) - \gamma\}\right)^{-1} < \infty.$$

Optimizing on $\gamma > \lambda(\beta)+\kappa(h)$, we get a lower bound which proves both variational formulas. $\qquad\square$

9.4   Tilt-Velocity Duality

Tilt-velocity duality is the familiar idea from large deviation theory that the macroscopic direction of the path is dual to an external field tilting the energy. In the positive temperature setting this is exactly the convex duality of the quenched large deviation principle for the endpoint of the path (see Remark 4.2 in [197]). Call a vector $h \in \mathbb{R}^d$ a tilt and elements $\xi \in \bar{\mathscr{B}}_1$ directions or velocities.

Equation (9.4) with P^h reads

$$p_{\mathrm{pl}}(\beta, h) = \sup_{\xi \in \bar{\mathscr{B}}_1} \{p_{\mathrm{pp}}(\beta, \xi) + h \cdot \xi\}, \qquad h \in \mathbb{R}^d. \qquad (9.31)$$

Relation (9.31) is a convex duality, with $p_{\mathrm{pl}}(\beta)$ the conjugate of the convex function $f = -p_{\mathrm{pp}}(\beta)$. More precisely, extending p_{pp} with the value $-\infty$ outside $\bar{\mathscr{B}}_1$, we can take the supremum in (9.31) to all $\xi \in \mathbb{R}^d$ without changing the value of the supremum. By double duality, the biconjugate is the largest lower-semicontinuous function f^{**} with $f^{**} \leq f$. Since $p_{\mathrm{pp}}(\beta)$ is continuous on $\mathscr{B}_1$, we have

$$p_{\mathrm{pp}}(\beta, \xi) = \inf_{h \in \mathbb{R}^d} \{p_{\mathrm{pl}}(\beta, h) - h \cdot \xi\}, \qquad \xi \in \mathscr{B}_1. \qquad (9.32)$$

Definition 9.3 Fix β. We declare $h \in \mathbb{R}^d$ and $\xi \in \mathscr{B}_1$ are dual to each other if

$$p_{\mathrm{pl}}(\beta, h) = p_{\mathrm{pp}}(\beta, \xi) + h \cdot \xi.$$

Then every $\xi \in \mathscr{B}_1$ has a dual $h \in \mathbb{R}^d$.

With these preliminaries we extend Theorem 9.4 to the point-to-point case.

Theorem 9.8 *We have the variational formula for* $\xi \in \mathcal{B}_1$:

$$p_{pp}(\beta, \xi) = \inf_{B \in \mathcal{K}} \mathbb{P}-\text{ess} \sup_{\omega} \ln P \left[e^{\beta \omega(o) - B(\omega, o, s_1) - h(B) \cdot \xi - h_0(B)} \right]. \tag{9.33}$$

A minimizing $B \in \mathcal{K}$ *exists for each* β *and each* $\xi \in \mathcal{B}_1$.

Proof Recall that B is a cocycle with $\mathfrak{h}(B) = \mathfrak{h}$ if and only if $F(\mathfrak{x}, \mathfrak{y}) = -B(\mathfrak{x}, \mathfrak{y}) - \mathfrak{h} \cdot (\mathfrak{y} - \mathfrak{x})$ is a centered cocycle. We write

$$K(V, F) = \mathbb{P}-\text{ess} \sup_{\omega} \ln P \left[e^{\beta V(\omega, S_1) + F(\omega, o, s_1)} \right],$$

with $V_h(\omega, z) = \omega(0, 0) + \beta^{-1} h \cdot z$. The right-hand side of the claim is equal to

$$\inf_{\mathfrak{h} \in \mathbb{R}^{1+d}} \inf_{B : \mathfrak{h}(B) = \mathfrak{h}} K(V_0, -B) - h \cdot \xi - h_0 = \inf_{\mathfrak{h} \in \mathbb{R}^{1+d}} \inf_{F \in \mathcal{K}_0} K(V_h, F) - h \cdot \xi$$

$$= \inf_{h \in \mathbb{R}^d} \{ p_{pl}(\beta, h) - h \cdot \xi \} \quad \text{(by Theorem 9.4)}$$

$$= p_{pp}(\beta, \xi)$$

by (9.32). $\qquad\qquad\qquad\qquad\qquad\qquad\qquad\qquad\qquad\qquad\qquad\qquad\qquad\square$

9.5 Busemann Functions and Minimizing Cocycles

In this section we explain how to obtain cocycles recovering the potential, from limits of gradients of free energy. Such limits are called Busemann functions, they come in two variants, point-to-point and point-to-level.

For $m \leq n, \mathfrak{x} = (m, x), \mathfrak{y} = (n, y) \in \mathcal{G}$ consider the free energy

$$G_{\mathfrak{x}, \mathfrak{y}}^{\beta} = \ln P \left[e^{\beta H_{m,n}} \mathbf{1}_{S_n = y} | S_m = x \right] \tag{9.34}$$

$$= \ln Z_{n-m, y-x}(\theta_{m,x}\omega, \beta),$$

recall the definition of $\hat{x}_n(\xi)$ from Sect. 9.1 and denote $\hat{\mathfrak{x}}_n(\xi) = (n, x_n(\xi))$.

Definition 9.4 The point-to-point Busemann function in the direction $\xi \in \mathcal{B}_1$ is defined as

$$B_{pp}^{\xi}(\mathfrak{x}, \mathfrak{y}) = \lim_{n \to \infty} \left[G_{\mathfrak{x}, \hat{\mathfrak{x}}_n(\xi) + \mathfrak{z}}^{\beta} - G_{\mathfrak{y}, \hat{\mathfrak{x}}_n(\xi) + \mathfrak{z}}^{\beta} \right] \tag{9.35}$$

for $\mathfrak{x}, \mathfrak{y} \in \mathcal{G}, \mathfrak{z} \in (\{1\} \times \mathcal{N}) \cup \{o\}$, provided that the limits a.s. exist.

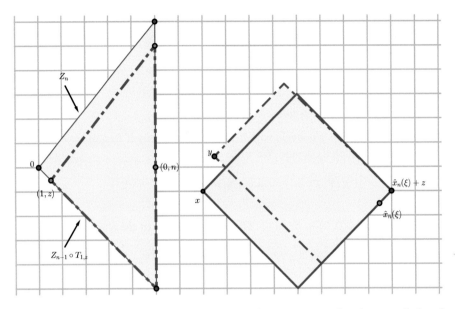

Fig. 9.1 Graphical representations of partition functions. Busemann functions are limits of logarithms of ratios of partition functions. A finite volume approximation of $\exp \beta B_{pl}^h(0, z)$ is shown on the *left* of the figure, and one of $\exp \beta B_{pp}^\xi(x, y)$ on the *right*

The extra step $\mathfrak{z}$ is introduced to check stationarity. We do not indicate in the notation the dependence in β (which is effective) and in $\mathfrak{z}$ (we will further assume the limit does not depend on $\mathfrak{z} = (1, z)$). Any point $\mathfrak{x} \in \mathscr{G}$ can reach $\hat{\mathfrak{x}}_n(\xi)$ with admissible steps when n is large enough.

Theorem 9.9 *Assume that for all* $\mathfrak{x}, \mathfrak{y} \in \mathscr{G}$ *and* $\mathbb{P}$-*a.e.* ω *the limits* (9.35) *exist for all* $\mathfrak{z} \in (\{1\} \times \mathscr{N}) \cup \{\mathfrak{o}\}$ *and do not depend on* $\mathfrak{z}$. *Then* $B_{pp}^\xi(\mathfrak{x}, \mathfrak{y})$ *is a* $L^1(\mathbb{P})$ *stationary cocycle that recovers* V_0. *Moreover,* $h(B_{pp}^\xi)$ *is dual to* ξ, *and*

$$p_{pp}(\beta, \xi) = -h(B_{pp}^\xi) \cdot \xi - h_0(B_{pp}^\xi).$$

The point of the theorem is that the Busemann function yields correctors for the variational formulas.

Now we turn to the second variant. (Both variants are represented in Fig. 9.1). Let $G_n^\beta(h)(\omega) = \ln Z_n(\omega, \beta; h)$, see (9.1).

Definition 9.5 Point-to-level Busemann functions are defined by a shift,

$$B_{pl}^h(\mathfrak{o}, \mathfrak{z}) = \lim_{n \to \infty} \left[G_n^\beta(h)(\omega) - G_{n-1}^\beta(h)(\theta_\mathfrak{z}\omega) \right], \tag{9.36}$$

for $\mathfrak{z} = (1, z), z \in \mathscr{N}$, when the limits a.s. exist.

Note that the above difference can be expressed in terms of the polymer measure

$$G_n^\beta(h)(\omega) - G_{n-1}^\beta(h)(\theta_\mathfrak{z}\omega) = \beta\omega(0,0) + h \cdot z + \ln P_{n-1}^{\beta,\omega}(S_1 = z). \tag{9.37}$$

Theorem 9.10 *Fix $h \in \mathbb{R}^d$, and assume that the limits (9.36) exist $\mathbb{P}$-a.s. for all $z \in \mathcal{N}$. Then $B_{pl}^h(\mathfrak{o},\mathfrak{z})$ extends to a L^1 stationary cocycle $B_{pl}^h(\mathfrak{x},\mathfrak{y})$ on $\mathcal{G}^2$, and cocycle $B_{pl}^h(\mathfrak{o},\mathfrak{z}) - h \cdot z$ recovers V_0. Moreover, $F(\omega,\mathfrak{x},\mathfrak{y}) = \mathfrak{h}(B_{pl}^h) \cdot (\mathfrak{x} - \mathfrak{y}) - B_{pl}^h(\mathfrak{x},\mathfrak{y})$ is a minimizer in (9.11) of $p_{pl}(\beta,h)$.*

Example 9.2 (Directed Polymer in Weak Disorder) The directed polymer in weak disorder illustrates Theorem 9.10. $\mathbb{P}$-almost surely for $z \in \mathcal{N}$,

$$\begin{aligned} \ln Z_n(\omega;\beta,h) - G_{n-1}^\beta(\theta_\mathfrak{z}\omega) &= \left[\ln W_n - \ln W_{n-1} \circ \theta_\mathfrak{z} + \lambda(\beta) + \kappa(h)\right] \\ &\longrightarrow \left[\ln W_\infty - \ln W_\infty \circ \theta_\mathfrak{z} + \lambda(\beta) + \kappa(h)\right] \\ &= -F(\mathfrak{o},\mathfrak{z}) + p_{pl}(\beta,h), \end{aligned}$$

with F defined by (9.26). Thus the Busemann function is

$$B_{pl}^h(\mathfrak{o},\mathfrak{z}) = -F(\mathfrak{o},\mathfrak{z}) + p_{pl}(\beta,h).$$

By Theorem 9.10, cocycle $B_{pl}^h(\mathfrak{o},\mathfrak{z}) - \mathfrak{h} \cdot \mathfrak{z}$ recovers V_0, as already observed in Example 9.1. The Busemann function recovers the corrector F identified in Example 9.1.

Proof of Theorem 9.9

Step 1: Cocycle B_{pp}^ξ recovers the potential V_0.
Using a first step decomposition and Markov property, definition (9.34) writes

$$e^{G_{\mathfrak{o},\hat{\mathfrak{x}}_n(\xi)+\mathfrak{z}}^\beta} = P\left[e^{\beta\omega(\mathfrak{o})}e^{G_{\mathfrak{s}_1,\hat{\mathfrak{x}}_n(\xi)+\mathfrak{z}}^\beta}\right],$$

or, equivalently,

$$1 = P\left[e^{\beta\omega(\mathfrak{o})}e^{G_{\mathfrak{s}_1,\hat{\mathfrak{x}}_n(\xi)+\mathfrak{z}}^\beta - G_{\mathfrak{o},\hat{\mathfrak{x}}_n(\xi)+\mathfrak{z}}^\beta}\right].$$

Taking the limit $n \to \infty$, we obtain the desired property,

$$1 = P\left[e^{\beta\omega(\mathfrak{o})}e^{B_{pp}^\xi(\mathfrak{o},\mathfrak{s}_1)}\right].$$

Step 2: B_{pp}^{ξ} is a stationary cocycle. For $\mathfrak{z} = (1, z), \mathfrak{z}' = (1, z')$ with $z, z' \in \mathcal{N}$, we have

$$
\begin{aligned}
B_{pp}^{\xi}(\mathfrak{x} + \mathfrak{z}, \mathfrak{y} + \mathfrak{z}) &= \lim_{n\to\infty} \left[G_{\mathfrak{x}+\mathfrak{z},\hat{\mathfrak{x}}_n(\xi)+\mathfrak{z}'}^{\beta} - G_{\mathfrak{y}+\mathfrak{z},\hat{\mathfrak{x}}_n(\xi)+\mathfrak{z}'}^{\beta} \right] \\
&= \lim_{n\to\infty} \left[G_{\mathfrak{x}+\mathfrak{z},\hat{\mathfrak{x}}_n(\xi)+\mathfrak{z}+\mathfrak{z}'}^{\beta} - G_{\mathfrak{y}+\mathfrak{z},\hat{\mathfrak{x}}_n(\xi)+\mathfrak{z}+\mathfrak{z}'}^{\beta} \right] \quad (9.38) \\
&= \lim_{n\to\infty} \left[G_{\mathfrak{x},\hat{\mathfrak{x}}_n(\xi)+\mathfrak{z}'}^{\beta} - G_{\mathfrak{y},\hat{\mathfrak{x}}_n(\xi)+\mathfrak{z}'}^{\beta} \right] \circ \theta_{\mathfrak{z}} \\
&= B_{pp}^{\xi}(\mathfrak{x}, \mathfrak{y}) \circ \theta_{\mathfrak{z}},
\end{aligned}
$$

where we used in (9.38) the assumption that the limit (9.35) does not depend on z. This yields stationarity. The cocycle property is straightforward: For $z' \in \mathcal{N}$,

$$
\begin{aligned}
B_{pp}^{\xi}(\mathfrak{x}, \mathfrak{y}) + B_{pp}^{\xi}(\mathfrak{y}, \mathfrak{z}) &= \lim_{n\to\infty} \left[G_{\mathfrak{x},\hat{\mathfrak{x}}_n(\xi)+\mathfrak{z}'}^{\beta} - G_{\mathfrak{y},\hat{\mathfrak{x}}_n(\xi)+\mathfrak{z}'}^{\beta} \right] \\
&\quad + \lim_{n\to\infty} \left[G_{\mathfrak{y},\hat{\mathfrak{x}}_n(\xi)+\mathfrak{z}'}^{\beta} - G_{\mathfrak{z},\hat{\mathfrak{x}}_n(\xi)+\mathfrak{z}'}^{\beta} \right] \\
&= \lim_{n\to\infty} \left[G_{\mathfrak{x},\hat{\mathfrak{x}}_n(\xi)+\mathfrak{z}'}^{\beta} - G_{\mathfrak{z},\hat{\mathfrak{x}}_n(\xi)+\mathfrak{z}'}^{\beta} \right] \\
&= B_{pp}^{\xi}(\mathfrak{x}, \mathfrak{z}).
\end{aligned}
$$

Step 3: Integrability of B_{pp}^{ξ}. Refer to Remark 5.3 in [118] and proof of Theorem 5.1 therein.

Step 4: Since cocycle B_{pp}^{ξ} recovers the potential V_0, we can apply (9.22) and we see that

$$
p_{pl}(\beta, h(B_{pp}^{\xi})) = -h_0(B_{pp}^{\xi}). \tag{9.39}
$$

Moreover, it is argued in [118] that convergence of L^1-norms in (9.35) holds, yielding the first equality below: for fixed m,

$$
\begin{aligned}
\mathbb{P}B_{pp}^{\xi}(\mathfrak{o}, \hat{\mathfrak{x}}_m(\xi)) &= \lim_{n\to\infty} \mathbb{P}\left[G_{\mathfrak{o},\hat{\mathfrak{x}}_n(\xi)}^{\beta} - G_{\hat{\mathfrak{x}}_m(\xi),\hat{\mathfrak{x}}_n(\xi)}^{\beta} \right] \\
&\overset{\text{Cesaro}}{=} \lim_{n\to\infty} \frac{1}{n} \sum_{k=m+1}^{n+m} \mathbb{P}\left[G_{\mathfrak{o},\hat{\mathfrak{x}}_k(\xi)}^{\beta} - G_{\hat{\mathfrak{x}}_m(\xi),\hat{\mathfrak{x}}_k(\xi)}^{\beta} \right] \\
&\overset{\text{stationarity}}{=} \lim_{n\to\infty} \frac{1}{n} \sum_{k=m+1}^{n+m} \mathbb{P}\left[G_{\mathfrak{o},\hat{\mathfrak{x}}_k(\xi)}^{\beta} - G_{\mathfrak{o},\hat{\mathfrak{x}}_k(\xi)-\hat{\mathfrak{x}}_m(\xi)}^{\beta} \right] \\
&= \lim_{n\to\infty} \frac{1}{n}\mathbb{P} \sum_{k=n+1}^{n+m} \left[G_{\mathfrak{o},\hat{\mathfrak{x}}_k(\xi)}^{\beta} - G_{\mathfrak{o},\hat{\mathfrak{x}}_{k-m}(\xi)}^{\beta} \right] + \mathcal{O}(1)
\end{aligned}
$$

$$\overset{\text{cancelation}}{=} \lim_{n\to\infty} \frac{1}{n}\mathbb{P}\left[\sum_{k=n+1}^{n+m} G^{\beta}_{0,\hat{\mathfrak{x}}_k(\xi)} - \sum_{k=1}^{m} G^{\beta}_{0,\hat{\mathfrak{x}}_k(\xi)}\right] + \mathcal{O}(1)$$

$$= m \lim_{n\to\infty} \frac{1}{n}\mathbb{P}G^{\beta}_{0,\hat{\mathfrak{x}}_n(\xi)} + \mathcal{O}(1).$$

By Remark 9.1 the limit in the right-hand side is equal to $p_{pp}(\beta, \xi)$. By the definition (9.9), the left-hand side is equal to $-\mathfrak{h}(B^{\xi}_{pp}) \cdot \mathfrak{x}_m(\xi)$. Dividing both sides by m and letting $m \to \infty$, we get

$$p_{pp}(\beta, \xi) = -\mathfrak{h}(B^{\xi}_{pp}) \cdot (1, \xi)$$

$$= -h(B^{\xi}_{pp}) \cdot \xi - h_0(B^{\xi}_{pp})$$

$$\overset{(9.39)}{=} p_{pl}(\beta, h(B^{\xi}_{pp})) - h(B^{\xi}_{pp}) \cdot \xi,$$

showing that $h(B^{\xi}_{pp})$ and ξ are dual. $\qquad\square$

Proof Sketch of proof of Theorem 9.10.

Step 1: Limits in (9.36) define B^h_{pl} as a cocycle: Fix $\mathfrak{x}, \mathfrak{y} \in \mathscr{G}$ and an admissible path $(x_i)_{i=0}^{\ell}$ from $\mathfrak{x} = (m, x_0)$ to $\mathfrak{y} = (m + \ell, y)$. Letting $z_i = x_i - x_{i-1}$, we write (taking $m = 0$ for simpler notations)

$$\sum_{i=0}^{\ell-1} B^h_{pl}(\theta_{i,x_i}\omega, z_{i+1}) = \lim_{n\to\infty}\sum_{i=0}^{\ell-1}[G^{\beta}_n(h)(\theta_{i,x_i}\omega) - G^{\beta}_{n-1}(h)(\theta_{i+1,x_{i+1}}\omega)]$$

$$= \lim_{n\to\infty}\sum_{i=0}^{\ell-1}[G^{\beta}_{n-i}(h)(\theta_{i,x_i}\omega) - G^{\beta}_{n-i-1}(h)(\theta_{i+1,x_{i+1}}\omega)]$$

$$\overset{\text{telescopic}}{=} \lim_{n\to\infty}[G^{\beta}_n(h)(\theta_{0,x_0}\omega) - G^{\beta}_{n-\ell}(h)(\theta_{\ell,x_\ell}\omega)]$$

$$= \lim_{n\to\infty}[G^{\beta}_n(h) - G^{\beta}_{n-\ell}(h) \circ \theta_{\mathfrak{y}-\mathfrak{x}}](\theta_{0,x_0}\omega),$$

which depends only on the endpoints $\mathfrak{x}, \mathfrak{y}$ of the path.
Step 2: Cocycle B^h_{pl} recovers the potential V_0: Using a first step decomposition and Markov property in definition (9.36), we get

$$e^{\beta G^h_n} = P\left[e^{\beta\omega_0 + h\cdot S_1} e^{\beta G^h_{n-1}\circ\theta_{1,S_1}}\right].$$

Reorganizing the terms and taking the limit $n \to \infty$, we derive that

$$1 = P\left[e^{\beta\omega_0 + h\cdot S_1} e^{-\beta B^h_{pl}(0, S_1)}\right],$$

showing that cocycle $f(\omega, z) = B_{\mathrm{pl}}^h(\mathfrak{o}, \mathfrak{z}) - h \cdot z$ recovers potential V_0.

Step 3: Integrability issue. Refer to Remark 5.3 in [118] and proof of Theorem 5.2 therein. See [118] for the remaining properties. $\qquad\square$

9.6 Gibbs' Variational Formulas: Energy/Entropy Balance

In this section we derive variational formulas in terms of probability measures on the spaces $\Omega_\ell = \Omega \times \mathcal{N}^\ell$, $\ell = 0, 1, \ldots$. We allow the potential $V(\omega, z_1^\ell)$ to depend also on the ℓ next steps of the path. This generality does not add any complication, but instead, makes things clearer. We are interested mostly in $V(\omega) = \omega(0, 0)$ (where $\ell = 0$) and $V(\omega, z) = \omega(0, 0) + h \cdot z$ (where $\ell = 1$). We write $z_k^\ell = (z_i)_{k \leq i \leq \ell}$, and we will use the notations $Z_i = S_i - S_{i-1}$ (not to be confused with partition functions) and S_z for shifts (not to be confused with the polymer path).

We first introduce the natural invariance properties, then the entropy, and then we state the variational formulas.

For $z \in \mathcal{N}$, introduce the mapping $S_z : \Omega_\ell \to \Omega_\ell$,

$$S_z(\omega, z_1^\ell) = (\theta_{1,z_1}\omega, (z_2^\ell, z)).$$

For $\ell = 0$, $S_z = \theta_{1,z}$ acts on $\Omega_0 \equiv \Omega$.

Definition 9.6 A probability measure μ on Ω_ℓ is S-invariant if

$$E^\mu[\max_{z \in \mathcal{N}} f \circ S_z] \geq E^\mu[f] \qquad \forall f : \Omega_\ell \to \mathbb{R} \text{ bounded measurable.}$$

Denote by $\mathcal{M}_s(\Omega_\ell)$ the space of all such measures.

Remark 9.5 For $\ell \geq 1$ (but not equal to 0), μ on Ω_ℓ is S-invariant iff

$$(\omega, Z_1^{\ell-1}) \overset{\text{law}}{=} (\theta_{1,Z_1}\omega, Z_2^\ell) \quad \text{under } \mu. \tag{9.40}$$

In particular, for $\ell = 1$, the condition deals with the evolved environment: $\theta_{1,Z_1}\omega \overset{\text{law}}{=} \omega$ under μ. Recall that the environment seen from the particle $\omega_n = \theta_{1,S_1}\omega$ is a Markov process. When $\ell = 1$, the measure μ induces a unique law $\bar{\mu}$ on Ω^2, the joint law of $(\omega, \theta_{1,S_1}\omega)$. The condition of S-invariance is that $\bar{\mu}$ have same first and second marginals.

Proof Let μ be S-invariant on Ω_ℓ. If f is a function of $(\omega, z_1^{\ell-1})$ only, then $f \circ S_z$ does not depend on z, and by S-invariance, $E^\mu[f \circ S_{z_\ell}] \geq E^\mu[f]$. The same argument with $-f$ shows that this is in fact an equality and (9.40) follows.

Conversely, assume (9.40). For any bounded f on Ω_ℓ, considering the function $\max_z f(\omega, z_1^{\ell-1}, z)$ on $\Omega_{\ell-1}$ one gets

$$E^\mu\left[\max_z f(\theta_{1,z_1}\omega, z_2^\ell, z)\right] = E^\mu\left[\max_z f(\omega, z_1^{\ell-1}, z)\right].$$

But the RHS is not smaller than $E^\mu[f]$ and S-invariance of f is proved. $\square$

We now give the full characterization of S-invariant measures—see Definition 9.6—as projections of measures on a bigger space, invariant under a mapping that we will call S.

Given $\mu \in \mathcal{M}_1(\Omega_\ell)$ with $\ell \geq 1$, denote by $\mu_\ell(z_\ell|\omega, z_1^{\ell-1})$ the conditional law under μ of Z_ℓ given $(\omega, Z_1^{\ell-1})$. We use it to define a transition kernel q on Ω_ℓ transforming (ω, z_1^ℓ) into $(\theta_{1,z_1}\omega, z_2^\ell, z)$ with probability

$$q_z(\omega, z_1^\ell) \equiv q\big((\omega, z_1^\ell), (\theta_{1,z_1}\omega, z_2^\ell, z)\big) = \mu_\ell(z|\theta_{1,z_1}\omega, z_2^\ell). \tag{9.41}$$

The first member is simply a convenient notation.

Let $z_k^\infty = (z_i; k \leq i) \in \mathcal{N}^{[k,\infty)}$ denote an infinite sequence of steps. On the space $\Omega_\infty = \Omega \times \mathcal{N}^{[1,\infty)}$ introduce the shift S,

$$S(\omega, z_1^\infty) = (\theta_{1,z_1}\omega, z_2^\infty), \tag{9.42}$$

and $\mathcal{M}_s(\Omega_\infty)$ the set of probability measures with are invariant under this shift.

Proposition 9.1 *Let $\ell \geq 0$ and $\mu \in \mathcal{M}_1(\Omega_\ell)$. Then the following are equivalent:*

1. *μ is S-invariant in the sense of Definition 9.6.*
2. *$\ell \geq 1$ and μ is invariant under the kernel (9.41) (constructed from μ itself), or $\ell = 0$ and there exists a Markov kernel $\{q_z(\omega) = q(\omega, \theta_{1,z}\omega); z \in \mathcal{N}\}$ on Ω such that $\mu q = \mu$.*
3. *μ is the marginal on Ω_ℓ of some probability measure on Ω_∞ which is invariant under the shift S from (9.42).*

Proof See [118, Proposition 7.1]. $\square$

Consider a random walk with uniform jumps and the Markov chain given by the following pair: the environment seen from the walker together with the ℓ next jumps. The transition kernel on Ω_ℓ is

$$p(\omega, S_z\omega) = |\mathcal{N}|^{-1} = (2d)^{-1}, \qquad z \in \mathcal{N}, \omega = (\omega, z_1^\ell) \in \Omega_\ell. \tag{9.43}$$

We now define an entropy $\bar{H}(\mu)$ of measures $\mu \in \mathcal{M}_1(\Omega_\ell)$, associated to this Markov chain p and to the environment law $\mathbb{P}$. For a transition kernel $q(\omega, \cdot)$ on

Ω_ℓ supported by $\{S_z \omega\}_{z \in \mathcal{N}}$, consider the relative entropy [see (3.41)]

$$\mathcal{H}(\mu \times q | \mu \times p) = \int_{\Omega_\ell} \sum_{z \in \mathcal{N}} q(\omega, S_z\omega) \ln \frac{q(\omega, S_z\omega)}{p(\omega, S_z\omega)} d\mu(\omega).$$

Denoting by μ_0 the marginal of μ on Ω, define

$$\bar{H}(\mu) = \begin{cases} \inf\{\mathcal{H}(\mu \times q | \mu \times p); q : \mu q = \mu\}, & \mu_0 \ll \mathbb{P} \\ \infty & \text{otherwise,} \end{cases} \quad (9.44)$$

where the infimum is taken over Markov kernels q on Ω_ℓ that fixes μ, i.e., μ is an invariant measure for the chain q. It is well known that $\bar{H} : \Omega_\ell \to [0, \infty]$ is convex. Now, we can state the variational formulas.

Theorem 9.11 (Theorem 2.3 in [198], Theorem 5.3 in [197]) *Let* $\mathbb{P}$ *i.i.d.,* $V : \Omega_\ell \to \mathbb{R}$ *with* $\mathbb{P}|V(\omega, z_1^\ell)|^p < \infty$ *for some* $p > d$. *Then*

$$p_{\text{pl}}(\beta) = \sup \{\beta E^\mu[\min(V, c)] - \bar{H}(\mu); \mu \in \mathcal{M}_s(\Omega_\ell), c > 0\}. \quad (9.45)$$

For the point-to-point free energy

$$p_{\text{pp}}(\beta, \xi) = \sup \{\beta E^\mu[\min(V, c)] - \bar{H}(\mu); \mu \in \mathcal{M}_s(\Omega_\ell), E^\mu[S_1] = \xi, c > 0\}. \quad (9.46)$$

The truncation by c is introduced in order to cover also unbounded potentials. For bounded potentials, the RHS has the usual form of the Gibbs' variational formula. In general, the truncation can be removed, see Theorem 7.5 in [118],

$$p_{\text{pl}}(\beta) = \sup \{\beta E^\mu[V] - \bar{H}(\mu); \mu \in \mathcal{M}_s(\Omega_\ell), E^\mu[V^-] < \infty\}. \quad (9.47)$$

Proof See the above three references. □

Open Problem 9.12 *Using the Gibbs' variational formula or the one with cocycles, prove that* $p(\beta) < \lambda(\beta)$ *for all positive* β *in space dimension* $d = 1, 2$. *In addition, get quantitative estimates comparable to those in Theorem 6.3.*

9.7   Variational Formula with a Functional Order Parameter

At the very moment when we finish this monograph, the preprint [25] is been posted. Bates and Chatterjee obtain another variational formula for the free energy, in the spirit of the ones from spin glass models [52, 192, 227, 228] optimizing over a functional order parameter. They describe limits of endpoint distributions of the polymer by an abstract object, so-called partitioned subprobability measure: it covers multiple blobs carrying positive mass escaping to infinity in different directions, as well as infinitesimal mass scattered over large areas.

In this formalism, the free energy is expressed as the solution of a variational problem over (subprobability) measures which are fixed point for an evolution operator describing the one step dynamics. The measure plays the role of an order parameter, and is a high dimensional object.

A beautiful result from [25] is that the order parameter is the trivial measure in high temperature region, but it has mass 1 in the low temperature region. As mentioned in Remark 5.5, the authors of this paper prove that the endpoint distribution is asymptotically purely atomic in the latter case.

Appendix A
Toolbox for Random Medium and Gibbs Measures

> *Les méthodes sont les habitudes de l'esprit et les économies de la mémoire.*
>
> (Antoine Rivarol, circa 1780)

A.1 Superadditive Lemmas

A real sequence $(u_n; n \geq 1)$ is called superadditive if

$$u_{n+m} \geq u_n + u_m, \qquad n, m \geq 1 .$$

The following result is standard, e.g., [97, Lemma 3.1.3].

Lemma A.1 (Superadditive Lemma) *If $(u_n; n \geq 1)$ is a superadditive sequence, then*

$$\lim_{n \to \infty} \frac{u_n}{n} = \sup_{m \geq 1} \frac{u_m}{m} \in \mathbb{R} \cup \{+\infty\} .$$

Proof Fix $m \geq 1$, and let $M = \min\{u_\ell; \ell = 1, \ldots, m-1\} \wedge 0$. Then, decomposing any integer n as $n = km + \ell$ with $k = \lfloor \frac{n}{m} \rfloor$ and $\ell \in \{0, \ldots, m-1\}$, we have by superadditivity, and setting $u_0 = 0$,

$$\frac{u_n}{n} \geq k \frac{u_m}{m} + \frac{u_\ell}{n} \geq k \frac{u_m}{m} + \frac{M}{n}$$

Hence,

$$\liminf_{n \to \infty} \frac{u_n}{n} \geq \frac{u_m}{m} , \qquad m \geq 1 .$$

This shows one inequality. Since $\limsup_{n \to \infty} n^{-1} u_n \leq \sup_{m \geq 1} m^{-1} u_m$ is trivial, the lemma is proved. □

The previous result has a stochastic counterpart, called the subadditive ergodic theorem, due originally to Kingman.

© Springer International Publishing AG 2017
F. Comets, *Directed Polymers in Random Environments*, Lecture Notes in Mathematics 2175, DOI 10.1007/978-3-319-50487-2

Theorem A.1 (Subadditive Ergodic Theorem) *Let $(X_{m,n}; m \leq n)$ be doubly-indexed random sequence such that:*

1. $X_{0,0} = 0$, and $X_{0,n} \leq X_{0,m} + X_{m,n}$ for $0 \leq m \leq n$,
2. $(X_{nk,(n+1)k}; n \geq 0)$ is a stationary ergodic process for all $k \geq 1$,
3. $(X_{m,m+k}; k \geq 0) \overset{\text{law}}{=} (X_{m+1,m+k+1}; k \geq 0)$ for all m,
4. $\mathbb{E}X_{0,1}^+ < \infty$.

Then, we have

$$\lim_{n \to \infty} \frac{X_{0,n}}{n} = \inf_{n \geq 1} \frac{\mathbb{E}X_{0,n}}{n} \in [-\infty, +\infty) , \quad Q - a.s. \tag{A.1}$$

If the limit is finite, then the convergence also holds in L^1-norm.

Note that $\mathbb{E}X_{0,n} \in [-\infty, \infty)$ by assumptions (1), (2) and (4). Note also that $\lim_{n \to \infty} \frac{\mathbb{E}X_{0,n}}{n} = \inf_{n \geq 1} \frac{\mathbb{E}X_{0,n}}{n}$ by the previous lemma applied to the super-additive sequence $u_n = -\mathbb{E}X_{0,n}$.

Proof See Liggett [165, Theorem 2.6, p.277]; Steele [221] gives a nice and elementary conceptual proof. □

A.2 Concentration Inequalities for Martingales with Bounded Increments

Loosely, concentration inequalities state that a function of many independent random variables, which does not depend much on any of them, strongly concentrated around its typical value. This theory, which has a long history in probability, has been renewed by a novel paper of Talagrand [225] in the mid-nineties. For a complete survey on the concentration of measure phenomenon, we refer to [51, 162, 163]

Here we will take the straightforward approach by martingales, which yields weaker results but of the same nature.

We start with the famous Azuma's lemma, see e.g. [12]. Maurey popularized its use by deriving an isoperimetric inequality for the symmetric group [170], showing its potential to study normed spaces, see [172]. This was the first step in opening a wide field of applications. This lemma is simple but quite useful. It deals with martingales M_n with bounded increments. For such, Markov inequality implies that $M_n = \mathcal{O}(n^{1/2})$ in probability. Azuma's lemma states that the large deviation probabilities $\mathbb{P}(|M_n|/n \geq r)$ are sub-gaussian.

Lemma A.2 (Azuma's Lemma) *Let $M_k, 0 \leq k \leq n$ be a $(\mathscr{H}_k)_k$-martingale starting from $M_0 = 0$ such that*

$$|M_k - M_{k-1}| \leq 1,$$

for $k = 1, \ldots, n$. Then, for $\theta \in \mathbb{R}$, we have

$$\mathbb{E}(e^{\theta M_n}) \leq \exp n\theta^2/2,$$

and, for all $r \geq 0$,

$$\mathbb{P}(M_n \geq nr) \leq e^{-nr^2/2}, \qquad \mathbb{P}(M_n \leq -nr) \leq e^{-nr^2/2}.$$

Proof By assumption $\Delta M_{k-1} \overset{\text{def}}{=} M_k - M_{k-1}$ has absolute value smaller than 1. From the barycentric relation

$$\Delta M_k = \frac{1 + \Delta M_k}{2} \cdot 1 + \frac{1 - \Delta M_k}{2} \cdot (-1),$$

it follows by convexity of the function $u \mapsto \exp \theta u$ for $\theta \in \mathbb{R}$, that

$$e^{\theta \Delta M_k} \leq \frac{1 + \Delta M_k}{2} e^{\theta} + \frac{1 - \Delta M_k}{2} e^{-\theta},$$

and by the martingale property,[1]

$$\mathbb{E}^{\mathscr{H}_k} e^{\theta \Delta M_k} \leq \cosh \theta \leq e^{\theta^2/2}.$$

The last inequality comes from the identities $\cosh \theta := (e^{\theta} + e^{-\theta})/2$, $(\ln \cosh \theta)' = \tanh \theta$, and the elementary fact $|\tanh \theta| \leq |\theta|$, $\theta \in \mathbb{R}$. Hence, for $\theta \in \mathbb{R}$,

$$\mathbb{E}(e^{\theta M_n}) = \mathbb{E}\mathbb{E}^{\mathscr{H}_{n-1}} \prod_{k=0}^{n-1} e^{\theta \Delta M_k}$$

$$= \mathbb{E}\Big(\prod_{k=0}^{n-2} e^{\theta \Delta M_k} \Big) \mathbb{E}^{\mathscr{H}_{n-1}} e^{\theta \Delta M_{n-1}}$$

$$\leq \mathbb{E} \prod_{k=0}^{n-2} e^{\theta \Delta M_k} e^{\theta^2/2}$$

$$\leq \exp n\theta^2/2$$

---

[1] Notation $\mathbb{E}^{\mathscr{H}_k}$ is for conditional expectation

$$\mathbb{E}^{\mathscr{H}_k}[X] = \mathbb{E}[X | \mathscr{H}_k].$$

by induction. This is the first claim. From this and Markov inequality we obtain for $\theta \geq 0$,

$$e^{n\theta r}\mathbb{P}(M_n \geq nr) \leq \mathbb{E}\left(e^{\theta M_n}; \{M_n \geq nr\}\right) \qquad (\theta \geq 0)$$
$$\leq \mathbb{E}(e^{\theta M_n})$$
$$\leq \exp n\theta^2/2.$$

Hence,

$$\mathbb{P}(M_n \geq nr) \leq \exp -n\sup_{\theta \geq 0}(r\theta - \theta^2/2),$$

which yields the desired bound $\exp -nr^2/2$ by taking the optimal θ ($\theta = r$ is non negative when $r \geq 0$). The same bound for $\mathbb{P}(M_n \leq -nr)$ follows by considering the martingale $(-M_k; k \leq n)$. $\qquad\square$

Corollary A.1 *Let $(X_k)_{k\leq n}$ a sequence of independent random variables taking values in some measurable space $(E, \mathscr{E})$, and*

$$Y_n = f_n(X_1, \ldots, X_n)$$

with $f_n : E^n \to \mathbb{R}$ a measurable function. Assume that for $k = 1, \ldots n$, all $x_1, \ldots, x_n, y_k$ in E,

$$|f_n(x_1, \ldots, x_{k-1}, x_k, x_{k+1}, \ldots, x_n) - f_n(x_1, \ldots, x_{k-1}, y_k, x_{k+1}, \ldots, x_n)| \leq 1$$

Then, we have the subgaussian estimates: $\mathbb{E}(\exp \theta[Y_n - \mathbb{E}Y_n]) \leq \exp n\theta^2/2$ for all real θ, and, for all $r \geq 0$,

$$\mathbb{P}(|Y_n - \mathbb{E}Y_n| \geq nr) \leq 2e^{-nr^2/2}.$$

Proof Let $\mathscr{H}_k = \sigma(X_1, \ldots X_k)$, $\mathbb{E}^{\mathscr{H}_k}$ the corresponding conditional expectation and $M_k = \mathbb{E}^{\mathscr{H}_k}(Y_n) - \mathbb{E}(Y_n)$. Then, $(M_k, k \leq n)$ is a $(\mathscr{H}_k, k \leq n)$-martingale, with $M_n = Y_n - \mathbb{E}Y_n$. Moreover, by independence,

$$\Delta M_{k-1} \stackrel{\text{def}}{=} M_k - M_{k-1}$$
$$= \mathbb{E}^{\mathscr{H}_k}(Y_n) - \mathbb{E}^{\mathscr{H}_{k-1}}(Y_n)$$
$$= \mathbb{E}^{\mathscr{H}_k}[f_n(X_1, \ldots, X_{k-1}, X_k, X_{k+1}, \ldots, X_n)$$
$$-f_n(X_1, \ldots, X_{k-1}, Z_k, X_{k+1}, \ldots, X_n)]$$

with Z_k an independent copy of X_k. (Observe that the random variable Z_k is integrated out in the second term of the last member.) By assumption, we have

$$|\Delta M_k| \leq 1, \quad k = 0, \ldots n - 1.$$

Then, the corollary directly follows from Azuma's Lemma A.2. $\quad\square$

Exercise A.1 We consider the setup of the polymer model. Assume that the random variables ω's are bounded, $|\omega(t, x)| \leq K$ a.s. for some finite $K > 0$. Using Azuma's lemma, show that

$$\mathbb{P}\left(\exp\{\theta[p_n(\omega, \beta) - \mathbb{P}p_n(\omega, \beta)]\}\right) \leq \exp\{C_1\theta^2/n\} ,$$

and deduce that

$$\mathrm{Var}_{\mathbb{P}}(p_n(\omega, \beta)) \leq C_2/n ,$$
$$\mathbb{P}\left(|p_n(\omega, \beta) - \mathbb{P}p_n(\omega, \beta)|\right) \leq C_3/\sqrt{n} .$$

We mention that the above estimate is not optimal, see Corollary 3.3 in [163] for sharp constants. Asking bounded oscillation in all variables as above is quite strong an assumption. In the next section, we will see the famous Gaussian concentration inequalities. These two result cover already a large family of environments.

A.3 Gaussian Laws

Denote by $g(x) = (2\pi)^{-1/2} \exp -x^2/2$ the gaussian density. If X has density g,

$$\frac{x}{1 + x^2}g(x) \leq \mathbb{P}(X \geq x) \leq (\frac{1}{x} \wedge 1)g(x) , \quad x > 0 , \tag{A.2}$$

Proof For the upper bound, write

$$\mathbb{P}(X > x) = \int_x^{+\infty} g(y)dy \leq \int_x^{+\infty} \frac{y}{x}g(y)dy = \frac{1}{x}g(x),$$

and recall the Chernoff's bound $\mathbb{P}(X \geq x) \leq e^{-x^2/2}$. In the other direction, define $h(x) = \mathbb{P}(X > x) - \frac{x}{1+x^2}g(x)$, check that $h(+\infty) = 0$ and

$$h'(x) = \frac{-2}{(1 + x^2)^2}g(x) < 0,$$

so that $h(x) \geq 0$ for all x. $\quad\square$

A.3.1 Concentration

In this section, we recall the standard concentration property of the gaussian measure, and we apply it to prove Remark 2.1.

Theorem A.2 (Gaussian Lipschitz Concentration) *[65] Let $M \geq 1$ be an integer. We consider $\mathbb{R}^M$ equipped with the usual euclidian norm $\| \cdot \|$. If X is a standard gaussian vector on some probability space (with a probability measure P) and F is a C-lipschitzian function ($|F(x) - F(y)| \leq C\|x - y\|$) from $\mathbb{R}^M$ to $\mathbb{R}$ then*

$$\mathrm{P}(e^{\lambda(F(X)-\mathrm{P}(F(X)))}) \leq e^{\frac{C^2\lambda^2}{2}}. \tag{A.3}$$

Therefore, we have the following concentration result

$$\mathrm{P}(|F(X) - \mathrm{P}(F(X))| \geq r) \leq 2e^{-\frac{r^2}{2C^2}} \tag{A.4}$$

Proof We start with the case of $F \in \mathscr{C}_b^2$, where we give a proof using stochastic calculus. (There exist more pedestrian proofs, e.g., Theorem 1.3.4 in [226], but we do not resist to the pleasure of this nice demonstration of the power of stochastic calculus; see [1, Lemma 2.2].) In this case we have

$$C := \|F\|_{\mathrm{lip}} = \max\{\|\nabla F(x)\|; x \in \mathbb{R}^M\},$$

as can be seen from

$$\begin{aligned}
F(x) - F(y) &= \int_0^1 \frac{d}{dt} F(y + t(x - y))dt \\
&= \int_0^1 \nabla F(y + t(x - y)) \cdot (x - y)dt,
\end{aligned}$$

and $|\nabla F(y + t(x - y)) \cdot (x - y)| \leq C\|x - y\|$. Considering $F - \mathrm{P}F(X)$ instead of F, we can assume that $\mathrm{P}F(X) = 0$ without loss of generality. Let B a standard Brownian motion in $\mathbb{R}^M$, and denote by $\mathscr{H}_t$ the σ-algebra generated by $B(s), s \leq t$. Then, $F(B(1)) \overset{\text{law}}{=} F(X)$, and we will prove (A.3) with $F(X) - \mathrm{P}(F(X))$ replaced by $M(1) = F(B(1))$. Introduce

$$M(t) = \mathrm{E}^{\mathscr{H}_t} F(B(1)), \quad t \in [0, 1],$$

which, as a martingale with respect to the Brownian filtration, can be written as a stochastic integral, namely (A.6) below. In fact,

$$M(t) = \mathrm{E}^{\mathscr{H}_t} F(B(t) + B(1) - B(t)) = (\mathscr{P}_{1-t}F)(B(t)),$$

where $(\mathscr{P}_t; t \geq 0)$ denotes the heat semi-group,

$$(\mathscr{P}_t F)(x) = \mathbb{E}[F(x + B(t))] .$$

Let $G(t,x) = (\mathscr{P}_{1-t}F)(x)$, note that $G(0,0) = 0$ by assumption, and write Itô's formula,

$$M(t) = G(t, B(t)) \tag{A.5}$$
$$= \int_0^t \nabla G(s, B(s)) \cdot dB(s) + \int_0^t \left[\frac{1}{2}\Delta G(s, B(s)) + \frac{\partial}{\partial s}G(s, B(s)) \right] ds,$$

with $\Delta = \sum_{i \leq M} \frac{\partial^2}{\partial x_i^2}$ the Laplacian operator. By Kolmogorov's backward equation,

$$\frac{\partial}{\partial s}G(s,x) = -\frac{\partial}{\partial t}(\mathscr{P}_t F)(x)_{|t=1-s}$$
$$= -\frac{1}{2}\Delta(\mathscr{P}_t F)(x)_{|t=1-s}$$
$$= -\frac{1}{2}\Delta G(s,x),$$

meaning that the last integral in (A.5) vanishes. Furthermore, we can compute

$$\nabla G(s,x) = (\mathscr{P}_{1-t}\nabla F)(x) .$$

Finally, we obtain[2]

$$M(t) = \int_0^t (\mathscr{P}_{1-s}\nabla F)(B(s)) \cdot dB(s) , \tag{A.6}$$

where

$$\|(\mathscr{P}_{1-t}\nabla F)(x)\| \leq (\mathscr{P}_{1-t}\|\nabla F\|)(x) \leq C.$$

Introduce the exponential martingale of a stochastic integral (A.6), we have for λ real,

$$1 = \mathbb{E}\exp\{\lambda M(1) - \frac{\lambda^2}{2}\int_0^1 \|(\mathscr{P}_{1-s}\nabla F)(B(s))\|^2 ds\}$$
$$\geq \mathbb{E}\exp\{\lambda M(1) - \frac{C^2\lambda^2}{2}\},$$

---

[2]The representation formula (A.6) is a very simple case of the Haussmann-Clark-Ocone formula.

ending the proof of (A.3). The derivation of (A.4) from (A.3) is standard, and entirely similar to the proof of the second part of Lemma A.2. This ends the proof in the special case of $F \in \mathscr{C}_b^2$. The proof in the general case follows from this special case by convolution with a smooth function. $\square$

We can derive from the above theorem a concentration result for the free energy of the polymer model with a standard gaussian environment ($\omega(t, x) \sim \mathscr{N}(0, 1)$):

Corollary A.2 *If the environment ω is standard gaussian then for all $\lambda \geq 0$,*

$$\mathbb{P}\left(e^{\lambda(\ln Z_n(\omega,\beta) - \mathbb{P}\ln Z_n(\omega,\beta))}\right) \leq e^{\frac{\beta^2 \lambda^2 n}{2}}. \tag{A.7}$$

Proof As a function of the environment, $F(\omega) = \ln Z_n(\omega, \beta)$ is $\beta \sqrt{n}$-lipschitzian. Indeed,

$$F(\omega) = \ln P\left[\exp \beta \sum_{x \in \mathbb{R}^d} \sum_{t=1}^{n} \omega(t, x)\mathbf{1}\{S_t = x\}\right],$$

$$\frac{\partial F}{\partial \omega(t, x)}(\omega) = \beta P_n^{\beta,\omega}(S_t = x),$$

$$\|\nabla F(\omega)\|^2 = \beta^2 \sum_{t,x} P_n^{\beta,\omega}(S_t = x)^2$$

$$= \beta^2 \sum_{t=1}^{n} \left[P_n^{\beta,\omega}\right]^{\otimes 2}(S_t = \tilde{S}_t)$$

$$\leq \beta^2 n.$$

Hence, by Cauchy-Schwarz,

$$|F(\omega') - F(\omega)| = \left|\int_0^1 \nabla F(\omega + r(\omega' - \omega)) \cdot (\omega' - \omega)dr\right|$$

$$\leq \beta \sqrt{n}\|\omega' - \omega\|.$$

Therefore, the result is a direct application of (A.3). $\square$

This implies (2.10) in Remark 2.1.

A.3.2 Integration by Parts

This is another important technical tool for Gaussian analysis.

Lemma A.3 (Integration by Part Formula) *If X is centered normal, and f is a smooth numerical function which does not grow too fast at infinity, i.e.,*

$$\lim_{|x|\to\infty} f(x) \exp\{-x^2/(2EX^2)\} = 0,$$

then

$$E\Big(Xf(X)\Big) = E\Big(X^2\Big)E\Big(f'(X)\Big) \qquad (A.8)$$

Proof Put $\sigma^2 = E\Big(X^2\Big)$ and $g_\sigma(x) = (2\pi\sigma^2)^{-1/2}\exp -x^2/(2\sigma^2)$ the density of X. By a standard integration by part,

$$\int_{-\infty}^{+\infty} xf(x)g_\sigma(x)dx = \sigma^2 \int_{-\infty}^{+\infty} f'(x)g_\sigma(x)dx$$

$\square$

Corollary A.3 (Integration by Part Formula for Gaussian Vectors) *If $(X, X_1, \ldots, X_n)$ is a centered, gaussian vector, and F is a smooth numerical function on $\mathbb{R}^n$ which does not grow too fast at infinity (e.g., $\lim_{\|x\|\to\infty} F(x)\exp\{-ax^2\} = 0$ for all $a > 0$), then*

$$E\Big(XF(X_1,\ldots,X_n)\Big) = \sum_{i=1}^{n} E\Big(XX_i\Big)E\Big(\frac{\partial F}{\partial x_i}(X_1,\ldots,X_n)\Big) \qquad (A.9)$$

Proof Put $a_i = E\Big(XX_i\Big)/E\Big(X^2\Big)$, and

$$X_i' = X_i - a_i X .$$

Then, $(X, X_1', \ldots, X_n')$ is a gaussian vector with X orthogonal to X_i' for all i: X is independent of $(X_1', \ldots, X_n')$. Now write

$$E\Big(XF(X_1,\ldots,X_n)\Big) = E\Big(XF(X_1' + c_1X,\ldots,X_n' + c_nX)\Big) ,$$

and use the integration by part formula (A.9) for $(X_1', \ldots, X_n')$ fixed to get the result.

$\square$

A.3.3 Slepian's Lemma

Now we come to the core of this section: for gaussian vectors $(X_1, \ldots, X_n)$, $(Y_1, \ldots, Y_n)$, we can compare

$$EF(X_1, \ldots, X_n) \text{ and } EF(Y_1, \ldots, Y_n)$$

for some functions F provided some inequality holds on the covariance matrices.

Theorem A.3 (Slepian's Lemma) *Let* $(X_i)_{i \leq n}$, $(Y_i)_{i \leq n}$ *two gaussian vectors with mean 0, and* $F : \mathbb{R}^n \to \mathbb{R}$ *a smooth function which does not grow too fast at infinity. If*

$$E(X_i^2) = E(Y_i^2) , \qquad E(X_i X_j) \leq E(Y_i Y_j) \tag{A.10}$$

and

$$\frac{\partial^2 F}{\partial x_i \partial x_j} \geq 0 \tag{A.11}$$

for all $i \neq j$, *then*

$$EF(X_1, \ldots, X_n) \leq EF(Y_1, \ldots, Y_n) .$$

This is an elegant and powerful formulation by Kahane [147], building on an original idea of Slepian [217].

Proof Without loss of generality, we may assume that the two vectors $\mathbf{X} = (X_i)_{i \leq n}$ and $\mathbf{Y} = (Y_i)_{i \leq n}$ are independent one of each other. (But of course, we do not assume that the coordinates of $(X_i)_{i \leq n}$ are independent!) For $t \in [0, 1]$, we define the "linear" interpolation between X and Y in the Gaussian space,

$$\mathbf{Z}(t) = \sqrt{1 - t}\mathbf{X} + \sqrt{t}\mathbf{Y}$$

and

$$\Phi(t) = EF(\mathbf{Z}(t)) . \tag{A.12}$$

The function $\Phi(t)$ interpolates between $\Phi(0) = EF(\mathbf{X})$ and $\Phi(1) = EF(\mathbf{Y})$. If F is smooth, it is differentiable,

$$\Phi'(t) = \frac{1}{2} \sum_{i=1}^{n} E\left[\left(\frac{1}{\sqrt{t}}Y_i - \frac{1}{\sqrt{1-t}}X_i\right)\frac{\partial F}{\partial x_i}(\mathbf{Z}(t))\right]$$

We now apply the integration by part formula (A.3) to each term in the sum:

$$E\left[\left(\frac{1}{\sqrt{t}}Y_i - \frac{1}{\sqrt{1-t}}X_i\right)\frac{\partial F}{\partial x_i}(\mathbf{Z}(t))\right] = \sum_{j=1}^{n} a_{i,j}E\left[\frac{\partial^2 F}{\partial x_i \partial x_j}(\mathbf{Z}(t))\right] ,$$

with

$$a_{i,j} = E\left[\left(\frac{1}{\sqrt{t}}Y_i - \frac{1}{\sqrt{1-t}}X_i\right)\left(\sqrt{1-t}X_j + \sqrt{t}Y_j\right)\right]$$
$$= E\left[Y_iY_j - X_iX_j\right]$$

by independence of $\mathbf{X}$ and $\mathbf{Y}$. Hence, from our assumptions $\Phi'(t) \geq 0$ for all $t \in (0,1)$, and $\Phi(0) \leq \Phi(1)$. The proof is complete. $\qquad\square$

A.4 Correlation Inequalities

Recall that a function $f : \mathbb{R}^k \to \mathbb{R}$ is increasing if $f(x) \leq f(y)$ whenever it $x_i \leq y_i \ \forall i \leq k$. This is equivalent to f being coordinatewise increasing, i.e., $f(x) \leq f(y)$ for all y such that $y_i = x_i$ for all i,j with $i \neq j$ and $x_j < y_j$.

Definition A.1 A family $X = (X_i; 1 \leq i \leq k)$ of real random variables defined on the same probability space are called **positively associated** if for any $f,g : \mathbb{R}^k \to \mathbb{R}$ bounded increasing functions,

$$\mathbb{E}[f(X)g(X)] \geq [\mathbb{E}f(X)][\mathbb{E}g(X)] . \tag{A.13}$$

The inequality (A.13) is called the Fortuyn-Kasteleyn-Ginibre (FKG) inequality. The inequality simply means that increasing functions are positively correlated. The following example is crucial in the applications, it is due to Harris [134].

Proposition A.1 (FKG-Harris Inequality) *A family of independent, real random variables is positively associated.*

Proof We prove Proposition A.1 by induction on k. For $k = 1$, consider an independent copy Y_1 of X_1, and write

$$\mathrm{Cov}(f(X_1), g(X_1)) = \mathbb{E}[f(X_1)g(X_1)] - [\mathbb{E}f(X_1)][\mathbb{E}g(X_1)]$$
$$= \frac{1}{2}\mathbb{E}\left[[f(X_1) - f(Y_1)][g(X_1) - g(Y_1)]\right] ,$$

where the integrand is a.s. non-negative by monotonicity of f, g. Let now $k > 1$, and write the standard decomposition

$$\text{Cov}(f(X), g(X)) = \text{Cov}(\mathbb{E}[f(X)|X_1], \mathbb{E}[g(X)|X_1]) + \mathbb{E}\text{Cov}(f(X), g(X)|X_1),$$

with $\text{Cov}(U, V|X_1)$ the covariance of U, V given X_1. We show that both terms are non-negative. For f non-decreasing, $\mathbb{E}[f(X)|X_1]$ is itself a non-decreasing function of X_1, and the first term will be non-negative according to the case $k = 1$. Indeed, by independence, $\mathbb{E}[f(X)|X_1] = \psi(X_1)$ is given by $\psi(x_1) = \mathbb{E}f(x_1, X_2, \ldots, X_k)$, and then, monotonicity follows from

$$\psi(x_1) - \psi(x_1') = \mathbb{E}[f(x_1, X_2, \ldots, X_k) - f(x_1', X_2, \ldots, X_k)] \geq 0 \quad \text{if} \quad x_1 \geq x_1'.$$

Again by independence, the second term is equal to $\mathbb{E}\Phi(X_1)$, where

$$\Phi(x_1) = \text{Cov}(F_{x_1}(X_2, \ldots, X_k), G_{x_1}(X_2, \ldots, X_k)),$$

with F_{x_1}, G_{x_1} the partial functions f, g at x_1, e.g., $F_{x_1}(x_2, \ldots, x_k) = f((x_1, \ldots, x_k))$. Clearly F_{x_1}, G_{x_1} are non-decreasing for all x_1, and the conditional covariance

$$\text{Cov}(f(X), g(X)|X_1)$$

is a.s. non-negative by the induction assumption. $\qquad\square$

A.5 Local Limit Theorem for the Simple Random Walk

We collect some known facts on the simple random walk, that we use in the text.

Recall some notation: $n \leftrightarrow x$, for $n \in \mathbb{N}^*$ and $x \in \mathbb{Z}^d$ such that $P(S_n = x) > 0$, $q^{(n)}(x) = P(S_n = x)$, and $\bar{q}^{(n)}(x) = 2(d/2\pi n)^{d/2} \exp(-d|x|^2/2n)$. We state the standard local limit theorem for the simple random walk on $\mathbb{Z}^d$, see e.g. [161].

Theorem A.4

$$\sup_{n \leftrightarrow x} |q^{(n)}(x) - \bar{q}^{(n)}(x)| = \mathcal{O}(n^{-(1+d/2)}). \tag{A.14}$$

In particular,

$$\sup_{n,x} q^{(n)}(x) = \mathcal{O}(n^{-d/2}), \tag{A.15}$$

and, for all $A \in (0, \infty)$ there exists a $c > 0$ such that

$$\inf\{q^{(n)}(x); n \leftrightarrow x, |x| \leq An^{1/2}\} \geq Cn^{-d/2}. \tag{A.16}$$

The theorem has a straightforward corollary (e.g., Corollary 3.2 in [234]).

Corollary A.4 *For all $A \in (0, \infty)$ there exists a finite $C = C(A, d)$ such that for all $t \in [0, 1)$, all n and all nonnegative function f,*

$$\sup\{P^{\otimes 2}[f((S_k, \tilde{S}_k)_{k \leq nt})|S_n = \tilde{S}_n = x]; |x| \leq An^{1/2}, n \leftrightarrow x\} \tag{A.17}$$

$$\leq C(1-t)^{-d} P^{\otimes 2}[f((S_k, \tilde{S}_k)_{k \leq nt})].$$

We need another property, which can be proved using potential theory for discrete Markov chains (see Corollary 3.4 in [234]).

Proposition A.2 *Let $d \geq 3$ and $v : \mathbb{Z}^d \times \mathbb{Z}^d \to \mathbb{R}$. Assume that*

$$0 < \inf_{x,y \in \mathbb{Z}^d} P^x \otimes P^y\left(e^{\sum_{i=1}^{\infty} v(S_i, \tilde{S}_i)}\right) \leq \sup_{x,y \in \mathbb{Z}^d} P^x \otimes P^y\left(e^{\sum_{i=1}^{\infty} v(S_i, \tilde{S}_i)}\right) < \infty.$$

Then, there exists some finite constant C such that

$$\sup_{x \in \mathbb{Z}^d} P^{\otimes 2}\left[e^{\sum_{i=1}^{n} v(S_i, \tilde{S}_i)}|S_n = \tilde{S}_n = x\right] \leq C. \tag{A.18}$$

A.6 Perron-Frobenius Eigenvalue of a Nonnegative Matrix

For a nonnegative square matrix A of size n, the spectral radius is

$$\rho(A) = \min_{g > 0} \max_{\omega \in \Omega} \frac{(Ag)(\omega)}{g(\omega)}, \tag{A.19}$$

with $\Omega = \{1, \ldots, n\}$—playing the role of coordinate index of the vector g. The positivity condition on g means that $g(\omega) > 0$ for all ω.

Proof Taking g a principal eigenvector we get that $\rho(A)$ is not smaller than the right-hand side in (A.19). If the inequality was strict, we could find some $g > 0$ with

$$\max_{\omega \in \Omega} \frac{(Ag)(\omega)}{g(\omega)} \leq \rho(A) - \varepsilon,$$

and by positivity, we could iterate

$$(A^k g)(\omega) \leq (\rho(A) - \varepsilon)^k g(\omega)$$

contradicting $(A^k g)(\omega) \simeq \rho(A)^k$.

For g principal eigenvector, the ratio $Ag(\omega)/g(\omega)$ does not depend on ω, and is equal to $\rho(A)$. $\qquad\square$

We note that (A.19) is the simplified form of (9.12) when the environment is periodic. In this case, the environment can be constructed on a finite space Ω and $\mathbb{P}$ is the uniform measure on the shifts of the environment over one period, instead of being a product measure on an infinite product space.

A.7 Linear Differential Equations: Duhamel Form

Duhamel's formula is the classical superposition property of linear differential equations: The solution to a general inhomogeneous linear equation can be expressed as a superposition of the general solution of the homogeneous equation and a particular solution of the equation with the forcing term. The formula is quite elementary, but also very general. For instance, the solution u of

$$du/dt - Lu = F(t), \quad u(0) = u_0,$$

for some linear operator L and a source term F, is given by

$$u(t) = e^{tL} u_0 + \int_0^t e^{(t-s)L} F(s) \, ds , \qquad (A.20)$$

in large generality. (Note first that when $Lu = \lambda u$ is simply a multiplicative operator, it reduces to the well-known method of variation of constants.) Indeed, the function $w(t) = e^{-tL} u(t) - u_0$ solves

$$dw/dt = e^{-tL} F(t), \quad w(0) = 0.$$

By integration we compute w, and thus

$$u(t) = e^{tL} u_0 + e^{tL} \int_0^t e^{-sL} F(s) ds = e^{tL} u_0 + \int_0^t e^{(t-s)L} F(s) ds$$

by linearity, and since L, e^{tL} commute.

References

1. R.J. Adler, *An Introduction to Continuity, Extrema, and Related Topics for General Gaussian Processes.* Institute of Mathematical Statistics Lecture Notes—Monograph Series, vol. 12 (Institute of Mathematical Statistics, Hayward, 1990)
2. T. Alberts, J. Clark, Nested critical points for a directed polymer on a disordered diamond lattice (2016), https://arxiv.org/abs/1602.06629
3. T. Alberts, M. Ortgiese, The near-critical scaling window for directed polymers on disordered trees. Electron. J. Probab. **18**(19), 24 (2013)
4. T. Alberts, K. Khanin, J. Quastel, Intermediate disorder regime for directed polymers in dimension 1+1. Phys. Rev. Lett. **105**, 090603 (2010)
5. T. Alberts, K. Khanin, J. Quastel, The continuum directed random polymer. J. Stat. Phys. **154**(1–2), 305–326 (2014)
6. T. Alberts, K. Khanin, J. Quastel, The intermediate disorder regime for directed polymers in dimension $1 + 1$. Ann. Probab. **42**(3), 1212–1256 (2014)
7. T. Alberts, J. Clark, S. Kocic, The intermediate disorder regime for a directed polymer model on a hierarchical lattice (2015), https://arxiv.org/abs/1508.04791
8. S. Albeverio, X.Y. Zhou, A martingale approach to directed polymers in a random environment. J. Theor. Probab. **9**(1), 171–189 (1996)
9. D.J. Aldous, G.K. Eagleson et al., On mixing and stability of limit theorems. Ann. Probab. **6**(2), 325–331 (1978)
10. K.S. Alexander, Subgaussian rates of convergence of means in directed first passage percolation (2011), https://arxiv.org/abs/1101.1549
11. K.S. Alexander, N. Zygouras, Subgaussian concentration and rates of convergence in directed polymers. Electron. J. Probab. **18**(5), 28 (2013)
12. N. Alon, J. Spencer, *The Probabilistic Method* (Wiley, New York, 2015)
13. G. Amir, I. Corwin, J. Quastel, Probability distribution of the free energy of the continuum directed random polymer in $1 + 1$ dimensions. Commun. Pure Appl. Math. **64**(4), 466–537 (2011)
14. S.N. Armstrong, P.E. Souganidis, Stochastic homogenization of Hamilton-Jacobi and degenerate Bellman equations in unbounded environments. J. Math. Pures Appl. (9) **97**(5), 460–504 (2012)
15. A. Auffinger, W.-K. Chen, A duality principle in spin glasses (2016), http://arxiv.org/abs/1605.01716
16. A. Auffinger, M. Damron, The scaling relation $\chi = 2\xi - 1$ for directed polymers in a random environment. ALEA Lat. Am. J. Probab. Math. Stat. **10**(2), 857–880 (2013)

© Springer International Publishing AG 2017

F. Comets, *Directed Polymers in Random Environments*, Lecture Notes in Mathematics 2175, DOI 10.1007/978-3-319-50487-2

17. A. Auffinger, M. Damron, A simplified proof of the relation between scaling exponents in first-passage percolation. Ann. Probab. **42**(3), 1197–1211 (2014)
18. A. Auffinger, O. Louidor, Directed polymers in random environment with heavy tails. Commun. Pure Appl. Math. **64**, 183–204 (2011)
19. A. Auffinger, J. Baik, I. Corwin, Universality for directed polymers in thin rectangles (2012), https://arxiv.org/abs/1204.4445
20. A. Auffinger, M. Damron, J.T. Hanson, 50 years of first passage percolation (preprint, 2015), https://arxiv.org/abs/1511.03262
21. M. Balázs, F. Rassoul-Agha, T. Seppäläinen, The random average process and random walk in a space-time random environment in one dimension. Commun. Math. Phys. **266**(2), 499–545 (2006)
22. M. Balázs, J. Quastel, T. Seppäläinen, Fluctuation exponent of the KPZ/stochastic Burgers equation. J. Am. Math. Soc. **24**(3), 683–708 (2011)
23. G. Barraquand, Some integrable models in the KPZ universality class, Ph.D thesis, Université Paris Diderot–Paris 7, 2015
24. G. Barraquand, I. Corwin, Random-walk in beta-distributed random environment. Probab. Theory Relat. Fields (2016, to appear), https://arxiv.org/abs/1503.04117
25. E. Bates, S. Chatterjee, The endpoint distribution of directed polymers (2016), https://arxiv.org/abs/1612.03443
26. I. Ben-Ari, Large deviations for partition functions of directed polymers in an IID field. Ann. Inst. Henri Poincaré Probab. Stat. **45**(3), 770–792 (2009)
27. M. Benaïm, R. Rossignol, Exponential concentration for first passage percolation through modified Poincaré inequalities. Ann. Inst. Henri Poincaré Probab. Stat. **44**(3), 544–573 (2008)
28. I. Benjamini, G. Kalai, O. Schramm, First passage percolation has sublinear distance variance. Ann. Probab. **31**(4), 1970–1978 (2003)
29. Q. Berger, H. Lacoin, The high-temperature behavior for the directed polymer in dimension 1+ 2 (2015), arXiv preprint arXiv:1506.09055
30. Q. Berger, F.L. Toninelli, On the critical point of the random walk pinning model in dimension $d = 3$. Electron. J. Probab. **15**(21), 654–683 (2010)
31. P. Bertin, Very strong disorder for the parabolic Anderson model in low dimensions. Ind. Math. (N.S.) **26**(1), 50–63 (2015)
32. L. Bertini, G. Giacomin, Stochastic Burgers and KPZ equations from particle systems. Commun. Math. Phys. **183**(3), 571–607 (1997)
33. J. Bertoin, Splitting at the infimum and excursions in half-lines for random walks and Lévy processes. Stoch. Process. Appl. **47**(1), 17–35 (1993)
34. J. Bertoin, R.A. Doney, On conditioning a random walk to stay nonnegative. Ann. Probab. **22**(4), 2152–2167 (1994)
35. S. Bezerra, S. Tindel, F. Viens, Superdiffusivity for a Brownian polymer in a continuous Gaussian environment. Ann. Probab. **36**(5), 1642–1675 (2008)
36. S. Bezerra, S. Tindel, F. Viens, Superdiffusivity for a Brownian polymer in a continuous Gaussian environment. Ann. Probab. **36**(5), 1642–1675 (2008)
37. J. Biggins, Martingale convergence in the branching random walk. J. Appl. Probab. **14**, 25–37 (1977)
38. M. Birkner, A condition for weak disorder for directed polymers in random environment. Electron. Commun. Probab. **9**, 22–25 (electronic) (2004)
39. M. Birkner, R. Sun, Annealed vs quenched critical points for a random walk pinning model. Ann. Inst. Henri Poincaré Probab. Stat. **46**(2), 414–441 (2010)
40. M. Birkner, R. Sun, Disorder relevance for the random walk pinning model in dimension 3. Ann. Inst. Henri Poincaré Probab. Stat. **47**(1), 259–293 (2011)
41. M. Birkner, A. Greven, F. den Hollander, Quenched large deviation principle for words in a letter sequence. Probab. Theory Relat. Fields **148**(3–4), 403–456 (2010)
42. M. Birkner, A. Greven, F. den Hollander, Collision local time of transient random walks and intermediate phases in interacting stochastic systems. Electron. J. Probab. **16**(20), 552–586 (2011)

43. G. Biroli, J.-P. Bouchaud, M. Potters, Extreme value problems in random matrix theory and other disordered systems. J. Stat. Mech.: Theory Exp. **2007**(07), P07019 (2007)

44. T. Bodineau, J. Martin, A universality property for last-passage percolation paths close to the axis. Electron. Commun. Probab. **10**, 105–112 (electronic) (2005)

45. C. Boldrighini, R.A. Minlos, A. Pellegrinotti, Almost-sure central limit theorem for directed polymers and random corrections. Commun. Math. Phys. **189**(2), 533–557 (1997)

46. E. Bolthausen, A note on the diffusion of directed polymers in a random environment. Commun. Math. Phys. **123**(4), 529–534 (1989)

47. A. Borodin, I. Corwin, Macdonald processes. Probab. Theory Relat. Fields **158**(1–2), 225–400 (2014)

48. A. Borodin, L. Petrov, Integrable probability: from representation theory to Macdonald processes. Probab. Surv. **11**, 1–58 (2014)

49. A. Borodin, I. Corwin, D. Remenik, Log-gamma polymer free energy fluctuations via a Fredholm determinant identity. Commun. Math. Phys. **324**(1), 215–232 (2013)

50. A. Borodin, I. Corwin, P.L. Ferrari, Free energy fluctuations for directed polymers in random media in 1+ 1 dimension. Commun. Pure Appl. Math. **67**(7), 1129–1214 (2014)

51. S. Boucheron, G. Lugosi, P. Massart, *Concentration Inequalities. A Nonasymptotic Theory of Independence*, with a foreword by Michel Ledoux (Oxford University Press, Oxford, 2013)

52. A. Bovier, *Statistical Mechanics of Disordered Systems*. Cambridge Series in Statistical and Probabilistic Mathematics (Cambridge University Press, Cambridge, 2006). A mathematical perspective

53. A. Cadel, S. Tindel, F. Viens, Sharp asymptotics for the partition function of some continuous-time directed polymers. Potential Anal. **29**(2), 139–166 (2008)

54. P. Calabrese, P. Le Doussal, A. Rosso, Free-energy distribution of the directed polymer at high temperature. Europhys. Lett. **90**(2), 20002 (2010)

55. A. Camanes, P. Carmona, The critical temperature of a directed polymer in a random environment. Markov Process. Relat. Fields **15**(1), 105–116 (2009)

56. F. Caravenna, R. Sun, N. Zygouras, Polynomial chaos and scaling limits of disordered systems. J. Eur. Math. Soc. **19**(1), 1–65 (2017)

57. F. Caravenna, R. Sun, N. Zygouras, Universality in marginally relevant disordered systems (2015), http://arxiv.org/abs/1510.06287

58. P. Carmona, Y. Hu, On the partition function of a directed polymer in a Gaussian random environment. Probab. Theory Relat. Fields **124**(3), 431–457 (2002)

59. P. Carmona, Y. Hu, Fluctuation exponents and large deviations for directed polymers in a random environment. Stoch. Process. Appl. **112**(2), 285–308 (2004)

60. P. Carmona, Y. Hu, Strong disorder implies strong localization for directed polymers in a random environment. ALEA Lat. Am. J. Probab. Math. Stat. **2**, 217–229 (2006)

61. R.A. Carmona, S.A. Molchanov, Parabolic Anderson problem and intermittency. Mem. Am. Math. Soc. **108**(518), viii+125 (1994)

62. R. Carmona, L. Koralov, S. Molchanov, Asymptotics for the almost sure lyapunov exponent for the solution of the parabolic Anderson problem. Random Oper. Stoch. Equ. **9**(1), 77–86 (2001)

63. S. Chatterjee, The universal relation between scaling exponents in first-passage percolation. Ann. Math. (2) **177**(2), 663–697 (2013)

64. S. Chatterjee, *Superconcentration and Related Topics*. Springer Monographs in Mathematics (Springer, Cham, 2014)

65. B. Cirel'son, I. Ibragimov, V. Sudakov, Norms of Gaussian sample functions, in *Proceedings of the Third Japan-USSR Symposium on Probability Theory (Tashkent, 1975)*. Lecture Notes in Mathematics, vol. 550 (Springer, Berlin, 1976), pp. 20–41

66. P. Collet, F. Koukiou, Large deviations for multiplicative chaos. Commun. Math. Phys. **147**, 329–342 (1992)

67. F. Comets, Large deviation estimates for a conditional probability distribution. Applications to random interaction Gibbs measures. Probab. Theory Relat. Fields **80**(3), 407–432 (1989)

68. F. Comets, Weak disorder for low dimensional polymers: the model of stable laws. Markov Proc. Relat. Fields **13**, 681–696 (2007)
69. F. Comets, M. Cranston, Overlaps and pathwise localization in the Anderson polymer model. Stoch. Process. Appl. **123**(6), 2446–2471 (2013)
70. F. Comets, Q. Liu, Rate of convergence for polymers in a weak disorder (2016), preprint arXiv:1605.05108
71. F. Comets, V.-L. Nguyen, Localization in log-gamma polymers with boundaries. Probab. Theory Relat. Fields **166**(1–2), 429–461 (2016)
72. F. Comets, V. Vargas, Majorizing multiplicative cascades for directed polymers in random media. ALEA Lat. Am. J. Probab. Math. Stat. **2**, 267–277 (2006)
73. F. Comets, N. Yoshida, Some new results on Brownian directed polymers in random environment. RIMS Kokyuroku **1386**, 50–66 (2004)
74. F. Comets, N. Yoshida, Brownian directed polymers in random environment. Commun. Math. Phys. **254**(2), 257–287 (2005)
75. F. Comets, N. Yoshida, Directed polymers in random environment are diffusive at weak disorder. Ann. Probab. **34**, 1746–1770 (2006)
76. F. Comets, N. Yoshida, Branching random walks in space-time random environment: survival probability, global and local growth rates. J. Theor. Probab. **24**(3), 657–687 (2011)
77. F. Comets, N. Yoshida, Localization transition for polymers in Poissonian medium. Commun. Math. Phys. **323**(1), 417–447 (2013)
78. F. Comets, T. Shiga, N. Yoshida, Directed polymers in a random environment: path localization and strong disorder. Bernoulli **9**(4), 705–723 (2003)
79. F. Comets, T. Shiga, N. Yoshida, Probabilistic analysis of directed polymers in a random environment: a review, in *Stochastic Analysis on Large Scale Interacting Systems*. Advanced Studies in Pure Mathematics, vol. 39 (Mathematical Society of Japan, Tokyo, 2004), pp. 115–142
80. F. Comets, S. Popov, M. Vachkovskaia, The number of open paths in an oriented ρ-percolation model. J. Stat. Phys. **131**(2), 357–379 (2008)
81. J. Cook, B. Derrida, Polymers on disordered hierarchical lattices: a nonlinear combination of random variables. J. Stat. Phys. **57**(1–2), 89–139 (1989)
82. J. Cook, B. Derrida, Directed polymers in a random medium: $1/d$ expansion and the n-tree approximation. J. Phys. A **23**(9), 1523–1554 (1990)
83. I. Corwin, The Kardar-Parisi-Zhang equation and universality class. Random Matrices Theory Appl. **1**(1), 1130001, 76 pp. (2012)
84. I. Corwin, Kardar-Parisi-Zhang universality. Not. Am. Math. Soc. **63**(3), 230–239 (2016)
85. I. Corwin, J. Quastel, D. Remenik, Continuum statistics of the Airy$_2$ process. Commun. Math. Phys. **317**(2), 347–362 (2013)
86. I. Corwin, N. O'Connell, T. Seppäläinen, N. Zygouras et al., Tropical combinatorics and Whittaker functions. Duke Math. J. **163**(3), 513–563 (2014)
87. M. Cranston, T.S. Mountford, T. Shiga, Lyapunov exponents for the parabolic Anderson model. Acta Math. Univ. Comenian. (N.S.) **71**(2), 163–188 (2002)
88. M. Cranston, D. Gauthier, T.S. Mountford, On large deviations for the parabolic Anderson model. Probab. Theory Relat. Fields **147**(1–2), 349–378 (2010)
89. F. Delarue, Martin Hairer, l'équation de KPZ et les structures de régularité. Gaz. Math. **143**, 15–28 (2015)
90. F. Delarue, R. Diel, Rough paths and 1d sde with a time dependent distributional drift: application to polymers. Probab. Theory Relat. Fields **165**, 1–63 (2016)
91. A. Dembo, O. Zeitouni, *Large Deviations Techniques and Applications*. Applications of Mathematics (New York), vol. 38, 2nd edn. (Springer, New York, 1998)
92. F. den Hollander, *Random Polymers*. Lecture Notes in Mathematics, vol. 1974 (Springer, Berlin, 2009). Lectures from the 37th Probability Summer School held in Saint-Flour, 2007
93. B. Derrida, R.B. Griffiths, Directed polymers on disordered hierarchical lattices. Europhys. Lett. **8**(2), 111 (1989)

94. B. Derrida, H. Spohn, Polymers on disordered trees, spin glasses, and traveling waves. J. Stat. Phys. **51**(5–6), 817–840 (1988). New directions in statistical mechanics (Santa Barbara, CA, 1987)
95. B. Derrida, M.R. Evans, E.R. Speer, Mean field theory of directed polymers with random complex weights. Commun. Math. Phys. **156**(2), 221–244 (1993)
96. B. Derrida, G. Giacomin, H. Lacoin, F.L. Toninelli, Fractional moment bounds and disorder relevance for pinning models. Commun. Math. Phys. **287**(3), 867–887 (2009)
97. J.-D. Deuschel, D.W. Stroock, *Large Deviations*. Pure and Applied Mathematics, vol. 137 (Academic, Boston, MA, 1989)
98. P.S. Dey, N. Zygouras, High temperature limits for $(1 + 1)$-dimensional directed polymer with heavy-tailed disorder. Ann. Probab. **44**(6), 4006–4048 (2016)
99. J. Ding, R. Eldan, A. Zhai, On multiple peaks and moderate deviations for the supremum of a Gaussian field. Ann. Probab. **43**(6), 3468–3493 (2015)
100. R. L. Dobrushin, B.S. Nakhapetyan, Strong convexity of the pressure for lattice systems of classical statistical physics. Theor. Math. Phys. **20**(2), 782–790 (1974)
101. M. Draief, J. Mairesse, N. O'Connell, Queues, stores, and tableaux. J. Appl. Probab. **42**(4), 1145–1167 (2005)
102. R. Durrett, Oriented percolation in two dimensions. Ann. Probab. **12**(4), 999–1040 (1984)
103. R. Durrett, *Lecture Notes on Particle Systems and Percolation*. The Wadsworth and Brooks/Cole Statistics/Probability Series (Wadsworth and Brooks/Cole Advanced Books and Software, Pacific Grove, CA, 1988)
104. R. Durrett, *Probability: Theory and Examples*, 2nd edn. (Duxbury Press, Belmont, CA, 1996)
105. S.F. Edwards, D.R. Wilkinson, The surface statistics of a granular aggregate, in *Proceedings of the Royal Society of London A: Mathematical, Physical and Engineering Sciences*, vol. 381 (The Royal Society, London, 1982), pp. 17–31
106. D. Erhard, F. den Hollander, G. Maillard, The parabolic anderson model in a dynamic random environment: basic properties of the quenched lyapunov exponent. Ann. Inst. Henri Poincaré Probab. Stat. **50**, 1231–1275 (2014)
107. M.R. Evans, B. Derrida, Improved bounds for the transition temperature of directed polymers in a finite-dimensional random medium. J. Stat. Phys. **69**(1), 427–437 (1992)
108. P.L. Ferrari, From interacting particle systems to random matrices. J. Stat. Mech. Theory Exp. **10**, P10016, 15 (2010)
109. P.L. Ferrari, H. Spohn, Random growth models, in *The Oxford Handbook of Random Matrix Theory* (Oxford University Press, Oxford, 2011), pp. 782–801
110. D. Fisher, C. Henley, D.A. Huse, Huse, Henley, and Fisher respond. Phys. Rev. Lett. **55**(26), 2924 (1985)
111. J. Franchi, Chaos multiplicatif: un traitement simple et complet de la fonction de partition, in *Séminaire de Probabilités*. Lecture Notes in Mathematics, vol. xxix (Springer, Berlin, 1993), pp. 194–201
112. R. Fukushima, N. Yoshida, On exponential growth for a certain class of linear systems. ALEA Lat. Am. J. Probab. Math. Stat. **9**(2), 323–336 (2012)
113. T. Funaki, J. Quastel, KPZ equation, its renormalization and invariant measures. Stoch. Partial Differ. Equ. Anal. Comput. **3**(2), 159–220 (2015)
114. N. Gantert, Y. Peres, Z. Shi, The infinite valley for a recurrent random walk in random environment. Ann. Inst. Henri Poincaré Probab. Stat. **46**(2), 525–536 (2010)
115. J. Gärtner, W. König, The parabolic Anderson model, in *Interacting Stochastic Systems* (Springer, Berlin, 2005), pp. 153–179
116. N. Georgiou, T. Seppäläinen, Large deviation rate functions for the partition function in a log-gamma distributed random potential. Ann. Probab. **41**(6), 4248–4286 (2013)
117. N. Georgiou, F. Rassoul-Agha, T. Seppäläinen, A. Yilmaz, Ratios of partition functions for the log-gamma polymer. Ann. Probab. **43**(5), 2282–2331 (2015)
118. N. Georgiou, F. Rassoul-Agha, T. Seppäläinen, Variational formulas and cocycle solutions for directed polymer and percolation models. Commun. Math. Phys. **346**(2), 741–779 (2016)
119. G. Giacomin, *Random Polymer Models* (Imperial College Press, London, 2007)

120. G. Giacomin, *Disorder and Critical Phenomena Through Basic Probability Models: École d'Été de Probabilités de Saint-Flour XL–2010*. Lecture Notes in Mathematics, vol. 2025 (Springer, New York, 2011)
121. A. Golosov, Localization of random walks in one-dimensional random environments. Commun. Math. Phys. **92**(4), 491–506 (1984)
122. P. Gonçalves, M. Jara, Nonlinear fluctuations of weakly asymmetric interacting particle systems. Arch. Ration. Mech. Anal. **212**(2), 597–644 (2014)
123. B.T. Graham, Sublinear variance for directed last-passage percolation. J. Theor. Probab. **25**(3), 687–702 (2012)
124. R.B. Griffiths, D. Ruelle, Strict convexity ("continuity") of the pressure in lattice systems. Commun. Math. Phys. **23**(3), 169–175 (1971)
125. M. Gubinelli, N. Perkowski, KPZ reloaded (2015), http://arxiv.org/abs/1508.03877
126. T. Gueudre, P. Le Doussal, J.-P. Bouchaud, A. Rosso, Ground-state statistics of directed polymers with heavy-tailed disorder. Phys. Rev. E **91**(6), 062110 (2015)
127. A. Guionnet, Large deviations and stochastic calculus for large random matrices. Probab. Surv. **1**, 72–172 (2004)
128. M. Hairer, Solving the KPZ equation. Ann. Math. (2) **178**(2), 559–664 (2013)
129. M. Hairer, Solving the KPZ equation, in *XVIIth International Congress on Mathematical Physics* (World Scientific Publishing, Hackensack, NJ, 2014), p. 419
130. M. Hairer, A theory of regularity structures. Invent. Math. **198**(2), 269–504 (2014)
131. C.H. Hall, Martingale limit theory and its applications, in *Probability and Mathematical Statistics* (Academic, New York/London, 1980)
132. T. Halpin-Healy, K.A. Takeuchi, A KPZ cocktail—shaken, not stirred . . . toasting 30 years of kinetically roughened surfaces. J. Stat. Phys. **160**(4), 794–814 (2015)
133. B. Hambly, J.B. Martin, Heavy tails in last-passage percolation. Probab. Theory Relat. Fields **137**(1–2), 227–275 (2007)
134. T.E. Harris, A lower bound for the critical probability in a certain percolation process. Proc. Camb. Philos. Soc. **56**, 13–20 (1960)
135. J.M. Harrison, *Brownian Motion and Stochastic Flow Systems*. Wiley Series in Probability and Mathematical Statistics: Probability and Mathematical Statistics (Wiley, New York, 1985)
136. H. Heil, M. Nakashima, N. Yoshida, Branching random walks in random environment are diffusive in the regular growth phase. Electron. J. Probab. **16**(48), 1316–1340 (2011)
137. Y. Hu, Q.-M. Shao, A note on directed polymers in Gaussian environments. Electron. Commun. Probab. **14**, 518–528 (2009)
138. Y. Hu, N. Yoshida, Localization for branching random walks in random environment. Stoch. Process. Appl. **119**(5), 1632–1651 (2009)
139. D. Hundertmark, A short introduction to Anderson localization, in *Analysis and Stochastics of Growth Processes and Interface Models* (Oxford University Press, Oxford, 2008), pp. 194–218
140. D.A. Huse, C.L. Henley, Pinning and roughening of domain walls in ising systems due to random impurities. Phys. Rev. Lett. **54**(25), 2708 (1985)
141. J. Imbrie, T. Spencer, Diffusion of directed polymers in a random environment. J. Stat. Phys. **52**(3–4), 609–626 (1988)
142. D. Ioffe, Multidimensional random polymers: a renewal approach, in *Random Walks, Random Fields, and Disordered Systems* (Springer, Cham, 2015), pp. 147–210
143. S. Janson, *Gaussian Hilbert Spaces*, vol. 129 (Cambridge University Press, Cambridge, 1997)
144. K. Johansson, Shape fluctuations and random matrices. Commun. Math. Phys. **209**(2), 437–476 (2000)
145. K. Johansson, Discrete orthogonal polynomial ensembles and the Plancherel measure. Ann. Math. (2) **153**(1), 259–296 (2001)
146. K. Johansson, Random matrices and determinantal processes, in *Mathematical Statistical Physics* (Elsevier B.V., Amsterdam, 2006), pp. 1–55
147. J.-P. Kahane, Une inégalité du type de Slepian et Gordon sur les processus gaussiens. Isr. J. Math. **55**(1), 109–110 (1986)

148. J.-P. Kahane, J. Peyrière, Sur certaines martingales de Benoit Mandelbrot. Adv. Math. **22**(2), 131–145 (1976)
149. K. Kalbasi, T.S. Mountford, Feynman-Kac representation for the parabolic Anderson model driven by fractional noise. J. Funct. Anal. **269**(5), 1234–1263 (2015)
150. M. Kardar, Roughening by impurities at finite temperatures. Phys. Rev. Lett. **55**(26), 2923–2923 (1985)
151. M. Kardar, G. Parisi, Y.-C. Zhang, Dynamic scaling of growing interfaces. Phys. Rev. Lett. **56**(9), 889–892 (1986)
152. J. Kertesz, V.K. Horvath, F. Weber, Self-affine rupture lines in paper sheets. Fractals **1**(01), 67–74 (1993)
153. H. Kesten, *Percolation Theory for Mathematicians*. Progress in Probability and Statistics, vol. 2 (Birkhäuser, Boston, MA, 1982)
154. E. Kosygina, S.R.S. Varadhan, Homogenization of Hamilton-Jacobi-Bellman equations with respect to time-space shifts in a stationary ergodic medium. Commun. Pure Appl. Math. **61**(6), 816–847 (2008)
155. T. Kriecherbauer, J. Krug, A pedestrian's view on interacting particle systems, KPZ universality and random matrices. J. Phys. A **43**(40), 403001, 41 (2010)
156. J. Krug, H. Spohn, Kinetic roughening of growing surfaces, in *Solids Far from Equilibrium*, ed. by Godrèche (Cambridge University Press, Cambridge, 1992), pp. 477–582
157. H. Lacoin, New bounds for the free energy of directed polymers in dimension $1+1$ and $1+2$. Commun. Math. Phys. **294**(2), 471–503 (2010)
158. H. Lacoin, Influence of spatial correlation for directed polymers. Ann. Probab. **39**(1), 139–175 (2011)
159. H. Lacoin, G. Moreno, Directed polymers on hierarchical lattices with site disorder. Stoch. Process. Appl. **120**(4), 467–493 (2010)
160. A. Lagendijk, B. van Tiggelen, D.S. Wiersma, Fifty years of anderson localization. Phys. Today **62**(8), 24–29 (2009)
161. G.F. Lawler, *Intersections of Random Walks*. Modern Birkhäuser Classics (Birkhäuser/Springer, New York, 2013). Reprint of the 1996 edition
162. M. Ledoux, Isoperimetry and Gaussian analysis, in *Lectures on Probability Theory and Statistics (Saint-Flour, 1994)*. Lecture Notes in Mathematics, vol. 1648 (Springer, Berlin, 1996), pp. 165–294
163. M. Ledoux, Concentration of measure and logarithmic Sobolev inequalities, in *Séminaire de Probabilités, XXXIII*. Lecture Notes in Mathematics, vol. 1709 (Springer, Berlin, 1999), pp. 120–216
164. C. Licea, C.M. Newman, M.S.T. Piza, Superdiffusivity in first-passage percolation. Probab. Theory Relat. Fields **106**(4), 559–591 (1996)
165. T.M. Liggett, *Interacting Particle Systems*. Classics in Mathematics (Springer, Berlin, 2005). Reprint of the 1985 original
166. Q. Liu, On generalized multiplicative cascades. Stoch. Process. Appl. **86**(2), 263–286 (2000)
167. Q. Liu, F. Watbled, Exponential inequalities for martingales and asymptotic properties of the free energy of directed polymers in a random environment. Stoch. Process. Appl. **119**(10), 3101–3132 (2009)
168. B. Mandelbrot, Multiplications aléatoires itérées et distributions invariantes par moyenne pondérée aléatoire. C. R. Acad. Sci. Paris Sér. A **278**, 289–292 (1974)
169. B. Mandelbrot, Multiplications aléatoires itérées et distributions invariantes par moyenne pondérée aléatoire: quelques extensions. C. R. Acad. Sci. Paris Sér. A **278**, 355–358 (1974)
170. B. Maurey, Construction de suites symétriques. C. R. Acad. Sci. Paris Sér. A-B **288**(14), A679–A681 (1979)
171. O. Mejane, Upper bound of a volume exponent for directed polymers in a random environment. Ann. Inst. H. Poincaré Probab. Stat. **40**(3), 299–308 (2004)
172. V.D. Milman, G. Schechtman, *Asymptotic Theory of Finite-Dimensional Normed Spaces*. Lecture Notes in Mathematics, vol. 1200 (Springer, Berlin, 1986). With an appendix by M. Gromov

173. M. Miura, Y. Tawara, K. Tsuchida, Strong and weak disorder for lévy directed polymers in random environment. Stoch. Anal. Appl. **26**(5), 1000–1012 (2008)
174. C. Monthus, T. Garel, Freezing transition of the directed polymer in a 1+d random medium: location of the critical temperature and unusual critical properties. Phys. Rev. E **74**, 011101 (2016)
175. G. Moreno, Convergence of the law of the environment seen by the particle for directed polymers in random media in the L^2 region. J. Theor. Probab. **23**(2), 466–477 (2010)
176. G.R. Moreno Flores, On the (strict) positivity of solutions of the stochastic heat equation. Ann. Probab. **42**(4), 1635–1643 (2014)
177. G. Moreno Flores, J. Quastel, D. Remenik, Endpoint distribution of directed polymers in $1+1$ dimensions. Commun. Math. Phys. **317**(2), 363–380 (2013)
178. G.R. Moreno Flores, T. Seppäläinen, B. Valkó, Fluctuation exponents for directed polymers in the intermediate disorder regime. Electron. J. Probab. **19**(89), 28 (2014)
179. G. Moreno Flores, J. Quastel, D. Remenik, in progress
180. J. Moriarty, N. O'Connell, On the free energy of a directed polymer in a Brownian environment. Markov Process. Relat. Fields **13**(2), 251–266 (2007)
181. P. Mörters, M. Ortgiese, Minimal supporting subtrees for the free energy of polymers on disordered trees. J. Math. Phys. **49**(12), 125203, 21 (2008)
182. C. Mueller, On the support of solutions to the heat equation with noise. Stoch. Stoch. Rep. **37**(4), 225–245 (1991)
183. Y. Nagahata, N. Yoshida, Localization for a class of linear systems. Electron. J. Probab. **15**(20), 636–653 (2010)
184. M. Nakashima, A remark on the bound for the free energy of directed polymers in random environment in $1 + 2$ dimension. J. Math. Phys. **55**(9), 093304, 14 (2014)
185. J. Neveu, *Martingales à Temps Discret* (Masson et Cie, éditeurs, Paris, 1972)
186. C.M. Newman, *Topics in Disordered Systems*. Lectures in Mathematics ETH Zürich (Birkhäuser Verlag, Basel, 1997)
187. C.M. Newman, M.S.T. Piza, Divergence of shape fluctuations in two dimensions. Ann. Probab. **23**(3), 977–1005 (1995)
188. V.L. Nguyen, A note about domination and monotonicity in disordered systems (2016). arXiv preprint arXiv:1606.01835
189. V.-L. Nguyen, Polymères dirigés en milieu aléatoire: systèmes intégrables, ordres stochastiques, Ph.D thesis, Université Paris Diderot, 2016
190. N. O'Connell, M. Yor, Brownian analogues of Burke's theorem. Stoch. Process. Appl. **96**(2), 285–304 (2001)
191. P. Olsen, R. Song, Diffusion of directed polymers in a strong random environment. J. Stat. Phys. **83**(3–4), 727–738 (1996)
192. D. Panchenko, *The Sherrington-Kirkpatrick Model*. Springer Monographs in Mathematics (Springer, New York, 2013)
193. M. Petermann, Superdiffusivity of polymers in random environment, Ph.D thesis, Univ. Zürich, 2000
194. M.S.T. Piza, Directed polymers in a random environment: some results on fluctuations. J. Stat. Phys. **89**(3–4), 581–603 (1997)
195. M. Prähofer, H. Spohn, Scale invariance of the PNG droplet and the Airy process. J. Stat. Phys. **108**(5–6), 1071–1106 (2002). Dedicated to David Ruelle and Yasha Sinai on the occasion of their 65th birthdays.
196. J. Quastel, Introduction to KPZ, in *Current Developments in Mathematics, 2011* (International Press, Somerville, MA, 2012), pp. 125–194
197. F. Rassoul-Agha, T. Seppäläinen, Quenched point-to-point free energy for random walks in random potentials. Probab. Theory Relat. Fields **158**(3–4), 711–750 (2014)
198. F. Rassoul-Agha, T. Seppäläinen, A. Yilmaz, Quenched free energy and large deviations for random walks in random potentials. Commun. Pure Appl. Math. **66**(2), 202–244 (2013)
199. F. Rassoul-Agha, T. Seppäläinen, A. Yilmaz, Variational formulas and disorder regimes of random walks in random potentials. Bernoulli **23**(1), 405–431, 02 (2017)

200. A. Rényi, On stable sequences of events. Sankhyā: Indian J. Stat. Ser. A **25**, 293–302 (1963)
201. S.I. Resnick, *Extreme Values, Regular Variation and Point Processes* (Springer, New York, 2013)
202. G.A. Ritter, Growth of random walks conditioned to stay positive. Ann. Probab. **9**(4), 699–704 (1981)
203. J.M. Rosenbluth, Quenched large deviation for multidimensional random walk in random environment: a variational formula. ProQuest LLC, Ann Arbor, MI, thesis (Ph.D.)—New York University, 2006
204. W. Rudin, *Real and Complex Analysis*, 3rd edn. (McGraw-Hill Book Co., New York, 1987)
205. W. Rudin, *Functional Analysis*. International Series in Pure and Applied Mathematics, 2nd edn. (McGraw-Hill, Inc., New York, 1991)
206. T. Sasamoto, H. Spohn, Exact height distributions for the KPZ equation with narrow wedge initial condition. Nucl. Phys. B **834**(3), 523–542 (2010)
207. T. Sasamoto, H. Spohn, One-dimensional kardar-parisi-zhang equation: an exact solution and its universality. Phys. Rev. Lett. **104**(23), 230602 (2010)
208. G. Schehr, S. Majumdar, A. Comtet, J. Randon-Furling, Exact distribution of the maximal height of p vicious walkers. Phys. Rev. Lett. **101**(15), 150601, 4 (2008)
209. T. Seppäläinen, Large deviations for lattice systems. I. Parametrized independent fields. Probab. Theory Relat. Fields **96**(2), 241–260 (1993)
210. T. Seppäläinen, Current fluctuations for stochastic particle systems with drift in one spatial dimension. Ens. Matemáticos **18**, 1–81 (2010)
211. T. Seppäläinen, Scaling for a one-dimensional directed polymer with boundary conditions. Ann. Probab. **40**, 19–73 (2012)
212. Z. Shi, *Branching Random Walks*. Lecture Notes in Mathematics, vol. 2151 (Springer, Cham, 2015). Lecture notes from the 42nd Probability Summer School held in Saint Flour, 2012, École d'Été de Probabilités de Saint-Flour [Saint-Flour Probability Summer School]
213. Y. Shiozawa, Central limit theorem for branching Brownian motions in random environment. J. Stat. Phys. **136**(1), 145–163 (2009)
214. Y. Shiozawa, Localization for branching Brownian motions in random environment. Tohoku Math. J. (2) **61**(4), 483–497 (2009)
215. Y.G. Sinai, The limit behavior of a one-dimensional random walk in a random environment. Teor. Veroyatnost. i Primenen. **27**(2), 247–258 (1982)
216. Y.G. Sinai, A remark concerning random walks with random potentials. Fundam. Math. **147**(2), 173–180 (1995)
217. D. Slepian, The one-sided barrier problem for Gaussian noise. Bell Syst. Tech. J. **41**, 463–501 (1962)
218. R. Song, X.Y. Zhou, A remark on diffusion of directed polymers in random environments. J. Stat. Phys. **85**(1–2), 277–289 (1996)
219. F. Spitzer, *Principles of Random Walk*. Graduate Texts in Mathematics, vol. 34, 2nd edn. (Springer, New York-Heidelberg, 1976)
220. H. Spohn, KPZ scaling theory and the semidiscrete directed polymer model, in *Random Matrix Theory, Interacting Particle Systems, and Integrable Systems*. Mathematical Sciences Research Institute Publications, vol. 65 (Cambridge University Press, New York, 2014), pp. 483–493
221. J.M. Steele, Kingman's subadditive ergodic theorem. Ann. Inst. H. Poincaré Probab. Stat. **25**(1), 93–98 (1989)
222. A.-S. Sznitman, *Brownian Motion, Obstacles and Random Media*. Springer Monographs in Mathematics (Springer, Berlin, 1998)
223. A.-S. Sznitman, Milieux aléatoires et petites valeurs propres, in *Milieux Aléatoires*. Panoramas et Synthèses, vol. 12 (Société Mathématique de France, Paris, 2001), pp. 13–36
224. K.A. Takeuchi, M. Sano, Universal fluctuations of growing interfaces: evidence in turbulent liquid crystals. Phys. Rev. Lett. **104**(23), 230601 (2010)
225. M. Talagrand, A new look at independence. Ann. Probab. **24**(1), 1–34 (1996)

226. M. Talagrand, Mean field models for spin glasses: a first course. *Lectures on probability theory and statistics (Saint-Flour, 2000)*. Lecture Notes in Mathematics, vol. 1816 (Springer, Berlin, 2003), pp. 181–285

227. M. Talagrand, *Mean Field Models for Spin Glasses. Volume I. Basic Examples* (Springer, Heidelberg, 2011)

228. M. Talagrand, *Mean Field Models for Spin Glasses. Volume II. Advanced Replica-Symmetry and Low Temperature* (Springer, Heidelberg, 2011)

229. T. Thiery, Analytical methods and field theory for disordered systems, Ph.D thesis, École Normale Supérieure de Paris, 2016

230. D. Thouless, Anderson localization in the seventies and beyond. Int. J. Mod. Phys. B **24**, 1507–1525 (2010)

231. N. Torri, Pinning model with heavy tailed disorder. Stoch. Process. Appl. **126**(2), 542–571 (2016)

232. C.A. Tracy, H. Widom, Level-spacing distributions and the Airy kernel. Commun. Math. Phys. **159**(1), 151–174 (1994)

233. S.R.S. Varadhan, Large deviations for random walks in a random environment. Commun. Pure Appl. Math. **56**, 1222–1245 (2003)

234. V. Vargas, A local limit theorem for directed polymers in random media: the continuous and the discrete case. Ann. Inst. H. Poincaré Probab. Stat. **42**(5), 521–534 (2006)

235. V. Vargas, Strong localization and macroscopic atoms for directed polymers. Probab. Theory Relat. Fields **138**(3–4), 391–410 (2007)

236. F.G. Viens, T. Zhang, Almost sure exponential behavior of a directed polymer in a fractional Brownian environment. J. Funct. Anal. **255**(10), 2810–2860 (2008)

237. J.B. Walsh, An introduction to stochastic partial differential equations, in *École d'été de Probabilités de Saint-Flour, XIV—1984*. Lecture Notes in Mathematics, vol. 1180 (Springer, Berlin, 1986), pp. 265–439

238. D. Williams, Path decomposition and continuity of local time for one-dimensional diffusions. I. Proc. Lond. Math. Soc. (3) **28**, 738–768 (1974)

239. D. Williams, *Probability with Martingales*. Cambridge Mathematical Textbooks (Cambridge University Press, Cambridge, 1991)

240. M.V. Wüthrich, Scaling identity for crossing Brownian motion in a Poissonian potential. Probab. Theory Relat. Fields **112**(3), 299–319 (1998)

241. M.V. Wüthrich, Superdiffusive behavior of two-dimensional Brownian motion in a Poissonian potential. Ann. Probab. **26**(3), 1000–1015 (1998)

242. A. Yilmaz, Equality of averaged and quenched large deviations for random walks in random environments in dimensions four and higher. Probab. Theory Relat. Fields **149**(3–4), 463–491 (2011)

243. N. Yoshida, Central limit theorem for branching random walks in random environment. Ann. Appl. Probab. **18**(4), 1619–1635 (2008)

244. N. Yoshida, Localization for linear stochastic evolutions. J. Stat. Phys. **138**(4–5), 598–618 (2010)

245. N. Zygouras, Lyapounov norms for random walks in low disorder and dimension greater than three. Probab. Theory Relat. Fields **143**(3–4), 615–642 (2009)

Index

© Springer International Publishing AG 2017
F. Comets, *Directed Polymers in Random Environments*, Lecture Notes
in Mathematics 2175, DOI 10.1007/978-3-319-50487-2

LECTURE NOTES IN MATHEMATICS  Springer

Editors in Chief: J.-M. Morel, B. Teissier;

Editorial Policy

1. Lecture Notes aim to report new developments in all areas of mathematics and their applications – quickly, informally and at a high level. Mathematical texts analysing new developments in modelling and numerical simulation are welcome.

   Manuscripts should be reasonably self-contained and rounded off. Thus they may, and often will, present not only results of the author but also related work by other people. They may be based on specialised lecture courses. Furthermore, the manuscripts should provide sufficient motivation, examples and applications. This clearly distinguishes Lecture Notes from journal articles or technical reports which normally are very concise. Articles intended for a journal but too long to be accepted by most journals, usually do not have this "lecture notes" character. For similar reasons it is unusual for doctoral theses to be accepted for the Lecture Notes series, though habilitation theses may be appropriate.

2. Besides monographs, multi-author manuscripts resulting from SUMMER SCHOOLS or similar INTENSIVE COURSES are welcome, provided their objective was held to present an active mathematical topic to an audience at the beginning or intermediate graduate level (a list of participants should be provided).

   The resulting manuscript should not be just a collection of course notes, but should require advance planning and coordination among the main lecturers. The subject matter should dictate the structure of the book. This structure should be motivated and explained in a scientific introduction, and the notation, references, index and formulation of results should be, if possible, unified by the editors. Each contribution should have an abstract and an introduction referring to the other contributions. In other words, more preparatory work must go into a multi-authored volume than simply assembling a disparate collection of papers, communicated at the event.

3. Manuscripts should be submitted either online at www.editorialmanager.com/lnm to Springer's mathematics editorial in Heidelberg, or electronically to one of the series editors. Authors should be aware that incomplete or insufficiently close-to-final manuscripts almost always result in longer refereeing times and nevertheless unclear referees' recommendations, making further refereeing of a final draft necessary. The strict minimum amount of material that will be considered should include a detailed outline describing the planned contents of each chapter, a bibliography and several sample chapters. Parallel submission of a manuscript to another publisher while under consideration for LNM is not acceptable and can lead to rejection.

4. In general, **monographs** will be sent out to at least 2 external referees for evaluation.

   A final decision to publish can be made only on the basis of the complete manuscript, however a refereeing process leading to a preliminary decision can be based on a pre-final or incomplete manuscript.

   Volume Editors of **multi-author works** are expected to arrange for the refereeing, to the usual scientific standards, of the individual contributions. If the resulting reports can be forwarded to the LNM Editorial Board, this is very helpful. If no reports are forwarded or if other questions remain unclear in respect of homogeneity etc, the series editors may wish to consult external referees for an overall evaluation of the volume.

5. Manuscripts should in general be submitted in English. Final manuscripts should contain at least 100 pages of mathematical text and should always include

   – a table of contents;
   – an informative introduction, with adequate motivation and perhaps some historical remarks: it should be accessible to a reader not intimately familiar with the topic treated;
   – a subject index: as a rule this is genuinely helpful for the reader.
   – For evaluation purposes, manuscripts should be submitted as pdf files.

6. Careful preparation of the manuscripts will help keep production time short besides ensuring satisfactory appearance of the finished book in print and online. After acceptance of the manuscript authors will be asked to prepare the final LaTeX source files (see LaTeX templates online: https://www.springer.com/gb/authors-editors/book-authors-editors/manuscriptpreparation/5636) plus the corresponding pdf- or zipped ps-file. The LaTeX source files are essential for producing the full-text online version of the book, see http://link.springer.com/bookseries/304 for the existing online volumes of LNM). The technical production of a Lecture Notes volume takes approximately 12 weeks. Additional instructions, if necessary, are available on request from lnm@springer.com.

7. Authors receive a total of 30 free copies of their volume and free access to their book on SpringerLink, but no royalties. They are entitled to a discount of 33.3 % on the price of Springer books purchased for their personal use, if ordering directly from Springer.

8. Commitment to publish is made by a *Publishing Agreement*; contributing authors of multiauthor books are requested to sign a *Consent to Publish form*. Springer-Verlag registers the copyright for each volume. Authors are free to reuse material contained in their LNM volumes in later publications: a brief written (or e-mail) request for formal permission is sufficient.

Addresses:
Professor Jean-Michel Morel, CMLA, École Normale Supérieure de Cachan, France
E-mail: moreljeanmichel@gmail.com

Professor Bernard Teissier, Equipe Géométrie et Dynamique,
Institut de Mathématiques de Jussieu – Paris Rive Gauche, Paris, France
E-mail: bernard.teissier@imj-prg.fr

Springer: Ute McCrory, Mathematics, Heidelberg, Germany,
E-mail: lnm@springer.com

Printed in the United States
By Bookmasters